ELASTIC-PLASTIC FRACTURE MECHANICS TECHNOLOGY

Sponsored by
ASTM Committee E-24 on
Fracture Testing
through its
Subcommittee E24.06.02

ASTM SPECIAL TECHNICAL PUBLICATION 896
J. C. Newman, Jr., NASA Langley Research
Center,
and F. J. Loss, Materials Engineering Associates,
editors

ASTM Publication Code Number (PCN)
04-896000-30

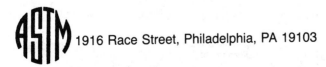 1916 Race Street, Philadelphia, PA 19103

Library of Congress Cataloging-in-Publication Data
Elastic-plastic fracture mechanics technology.
   (ASTM special technical publication; 896)
   Proceedings of a workshop.
   "ASTM publication code number (PCN) 04-896000-30."
   Includes bibliographies and index.
   1. Fracture mechanics—Congresses.
2. Elastoplasticity—Congresses.   I. Newman, J. C.   II. Loss, F. J.   III. ASTM
Committee E-24 on Fracture Testing.   Subcommittee E24.06.02.   IV. Series.
E24.06.02.   IV. Series.
TA409.E38   1986   620.1'126   85-22965
ISBN 0-8031-0449-9

NOTE

The Society is not responsible, as a body,
for the statements and opinions
advanced in this publication.

Printed in Baltimore, MD
December 1985

Foreword

This publication is the results of an ASTM Committee E24.06.02 Task Group round robin on fracture and a collection of papers presented at a workshop on Elastic-Plastic Fracture Mechanics Technology held at the regular Committee E-24 on Fracture Testing meeting in the Spring of 1983. The objective of the round robin and workshop was to evaluate and to document various elastic-plastic failure load prediction methods. J. C. Newman, Jr., NASA Langley Research Center, and F. J. Loss, Materials Engineering Associates, are editors of this publication.

Related
ASTM Publications

Elastic-Plastic Fracture Test Methods, STP 856 (1985), 04-856000-30

Elastic-Plastic Fracture: Second Symposium, Volume I: Inelastic Crack Analysis; Volume II: Fracture Curves and Engineering Applications, STP 803 (1983), Volume I—04-803001-30; Volume II—04-803002-30

Elastic-Plastic Fracture, STP 668 (1979), 04-668000-30

A Note of Appreciation
to Reviewers

The quality of the papers that appear in this publication reflects not only the obvious efforts of the authors but also the unheralded, though essential, work of the reviewers. On behalf of ASTM we acknowledge with appreciation their dedication to high professional standards and their sacrifice of time and effort.

ASTM Committee on Publications

ASTM Editorial Staff

Helen M. Hoersch
Janet R. Schroeder
Kathleen A. Greene
Bill Benzing

Contents

Introduction

Since the development of fracture mechanics, the materials scientists and design engineers have had an extremely useful concept with which to describe quantitatively the fracture behavior of solids. The use of fracture mechanics has permitted the materials scientists to conduct meaningful comparisons between materials on the influence of microstructure, stress state, and crack size on the fracture process. To the design engineer, fracture mechanics has provided a methodology to use laboratory fracture data (such as tests on compact specimens) to predict the fracture behavior of flawed structural components.

Many of the engineering applications of fracture mechanics have been centered around linear-elastic fracture mechanics (LEFM). This concept has proved to be invaluable for the analysis of brittle high-strength materials. LEFM concepts, however, become inappropriate when ductile low-strength materials are used. LEFM methods also become inadequate in the design and reliability analysis of many structural components. To meet this need, much experimental and analytical effort has been devoted to the development of elastic-plastic fracture mechanics (EPFM) concepts. Over the past two decades, many EPFM methods have been developed to assess the toughness of metallic materials and to predict failure of cracked structural components. However, for materials that exhibit large amounts of plasticity and stable crack growth prior to failure, there is no consensus of opinion on the most satisfactory method. To assess the accuracy and usefulness of many of these methods, an experimental and predictive round robin was conducted in 1979–1980 by Task Group E24.06.02 under the Applications Subcommittee of the ASTM Committee E-24 on Fracture Testing. The objective of the round robin was to verify experimentally whether the fracture analysis methods currently used could predict failure (maximum load or instability load) of complex structural components containing cracks from results of laboratory fracture toughness test specimens (such as the compact specimen) for commonly used engineering materials and thicknesses.

The ASTM Task Group E24.06.02 had also undertaken the task of organizing the documentation of various elastic-plastic fracture mechanics methods to assess flawed structural component behavior. The task group co-chairmen asked for the participation of interested members and, thus, six groups representing different methods were formed. These groups and corresponding chairmen were: (1) K_R-Resistance Curve Method, Chairmen D. E. McCabe and K. H. Schwalbe; (2)

Deformation Plasticity Failure Assessment Diagram (R-6), Chairman J. M. Bloom; (3) Dugdale Strip Yield Model with K_R-Resistance Curve Method, Chairman R. deWit, which is Appendix X of the first paper in this publication; (4) J_R-Resistance Curve Method, Chairmen H. A. Ernst and J. D. Landes; and (5) Crack-Tip-Opening Displacement (CTOD/CTOA) Approach, Chairman J. C. Newman, Jr. The chairmen were assigned the task of producing a written document explaining in detail a particular method following a common outline. The major objectives of these documents were to explain what laboratory tests were needed to determine the appropriate fracture parameter(s) and to demonstrate how the method is used to predict failure of cracked structural components.

J. C. Newman, Jr.

NASA Langley Research Center, Hampton, VA 23665; task group co-chairman and editor.

F. J. Loss

Materials Engineering Associates, Inc., Lanham, MD 20706; task group co-chairman and editor.

Experimental and Predictive
Round Robin

J. C. Newman, Jr.[1]

An Evaluation of Fracture Analysis Methods*

REFERENCE: Newman, J. C., Jr., **"An Evaluation of Fracture Analysis Methods,"** *Elastic-Plastic Fracture Mechanics Technology, ASTM STP 896,* J. C. Newman, Jr., and F. J. Loss, Eds., American Society for Testing and Materials, Philadelphia, 1985, pp. 5–96.

ABSTRACT: This paper presents the results of an experimental and predictive round robin conducted by the American Society for Testing and Materials (ASTM) Task Group E24.06.02 on Application of Fracture Analysis Methods. The objective of the round robin was to verify whether fracture analysis methods currently used can or cannot predict failure loads on complex structural components containing cracks. Fracture results from tests on compact specimens were used to make these predictions. Results of fracture tests conducted on various-size compact specimens made of 7075-T651 aluminum alloy, 2024-T351 aluminum alloy, and 304 stainless steel were supplied as baseline data to 18 participants. These participants used 13 different fracture analysis methods to predict failure loads on other compact specimens, middle-crack tension (formerly center-crack tension) specimens, and structurally configured specimens. The structurally configured specimen, containing three circular holes with a crack emanating from one of the holes, was subjected to tensile loading.

The accuracy of the prediction methods was judged by the variations in the ratio of predicted-to-experimental failure loads, and the prediction methods were ranked in order of minimum standard error. The range of applicability of the prediction methods was also considered in assessing their usefulness. For 7075-T651 aluminum alloy, the best methods (predictions within ±20% of experimental failure loads) were: the effective K_R-curve, the critical crack-tip-opening displacement (CTOD) criterion using a finite-element analysis, and the K_R-curve with the Dugdale model. For the 2024-T351 aluminum alloy, the best methods were: the Two-Parameter Fracture Criterion (TPFC), the CTOD criterion using the finite-element analysis, the K_R-curve with the Dugdale model, the Deformation Plasticity Failure Assessment Diagram (DPFAD), and the effective K_R-curve with a limit-load condition. For 304 stainless steel, the best methods were: limit-load (or plastic collapse) analyses, the CTOD criterion using the finite-element analysis, the TPFC, and the DPFAD. The failure loads were unknown to all participants except the author, who used both the TPFC and the CTOD criterion (finite-element analysis).

KEY WORDS: fracture (materials), elastic-plastic fracture, ductile fracture, tearing, stable crack growth, instability, stress-intensity factor, finite-element method, Dugdale model, J-integral, fracture criteria, elasticity, plasticity

[1] Senior scientist, NASA Langley Research Center, Hampton, VA 23665.

* The 17 Appendices to this paper were provided by individual contributers, as noted in the byline to each Appendix. A list of the participants, along with their affiliations, is also given in Table 1.

Nomenclature

A Area under load-displacement record, kN/mm

A_i Coefficients in residual strength equation (Eq 53)

A_{ij} Coefficients in stress-intensity factor equation (Eq 21)

A_{net} Net-section area on crack plane, mm^2

a Physical crack length (see Fig. 1), mm

a_e Effective crack length $(a + r_p)$, mm

a_0 Initial crack length, mm

a_u Crack length used in Theory of Ductile Fracture, mm

B Specimen thickness, mm

b Distance from small hole to edge of plate in three-hole-crack tension specimen, mm

c_i Coefficients in K_R-curve equation (Eq 14)

D Diameter of small hole in three-hole-crack tension specimen, mm

d Finite-element size in crack-tip region, mm

E Modulus of elasticity, MN/m^2

F Boundary-correction factor

J_{Ie} Elastic J-integral (K^2/E), kN/m

J_R Crack-growth resistance in terms of J-integral, kN/m

K Elastic stress-intensity factor, MN/m$^{3/2}$

K_c Critical (plastic-zone corrected) stress-intensity factor, MN/m$^{3/2}$

K_e Effective stress-intensity factor (Eq 45), MN/m$^{3/2}$

K_F Elastic-plastic fracture toughness from TPFC, MN/m$^{3/2}$

K_f Fracture toughness from three-dimensional finite-element analysis, MN/m$^{3/2}$

K_{Ic} Fracture toughness from "standard" ASTM Test Method for Plane-Strain Fracture Toughness of Metallic Materials (E 399-83) specimen, MN/m$^{3/2}$

K_{Icd} Fracture toughness used in Equivalent Energy method, MN/m$^{3/2}$

K_{Ie} Elastic stress-intensity factor at failure, MN/m$^{3/2}$

K_R Crack-growth resistance in terms of K, MN/m$^{3/2}$

K_r Ratio of stress-intensity factor to fracture toughness for Failure Assessment Diagram

M Number of predictions used in computing standard error

m Fracture toughness parameter from TPFC

N Nominal stress conversion factor (S/S_n)

n Ramberg-Osgood strain-hardening power

P Load, kN

P_c Calculated failure load, kN

P_f Experimental failure load, kN

P_p Predicted failure load, kN

P_L Plastic-collapse or limit load, kN

r_p Irwin's plastic-zone size, mm

S Gross-section stress, MN/m^2

S_L Plastic-collapse or limit stress, MN/m^2

S_n Nominal (net-section) stress, MN/m^2

S_r Ratio of applied stress to net section collapse stress for Failure Assessment Diagram

S_u Plastic-collapse (nominal) stress, MN/m^2

V_o Crack-mouth opening displacement, mm

V_{LL} Crack load-line displacement, mm

W Specimen width, mm

β Constraint factor (see Appendix X)

δ_c Critical CTOD from finite-element analysis, mm

Δa_e Effective crack extension, mm

Δa_p Physical crack extension, mm

ϵ Engineering strain

κ Ramberg-Osgood strain-hardening coefficient, MN/m^2

ρ Plastic-zone size, mm

σ Engineering stress, MN/m^2

σ_o Effective flow stress, MN/m^2

σ_u Ultimate tensile strength, MN/m^2

σ_{ys} Yield stress (0.2% offset), MN/m^2

σ_{yy} Normal stress acting in y-direction, MN/m^2

λ Crack aspect ratio (a/W)

ω Parameter used in Theory of Ductile Fracture

Subscripts

o   Denotes quantity determined from crack-mouth displacements

LL   Denotes quantity determined from load-line displacements

v Denotes quantity determined from visual measurements

Over the past two decades, many fracture analysis methods have been developed to assess the toughness of a metallic material and to predict failure of cracked structural components. For materials that fail under brittle conditions (small plastic-zone-to-plate-thickness ratios), the method based on linear-elastic fracture mechanics (LEFM), namely plane-strain fracture toughness (K_{Ic}), is widely accepted [1]. However, for materials that exhibit large amounts of plasticity and stable crack growth prior to failure, there is no consensus of opinion on the most satisfactory method. In recent years, a large number of elastic-plastic fracture mechanics methods have been developed [2–4]. To assess the accuracy and usefulness of many of these methods, an experimental and predictive round robin was conducted in 1979–1980 by Task Group E24.06.02 under the Applications Subcommittee of the American Society for Testing and Materials (ASTM) Committee E24 on Fracture Testing of Materials. The objective of the round robin was to determine whether the fracture analysis methods currently used can or cannot predict failure (maximum load or instability load) of complex structural components containing cracks from results of laboratory fracture toughness test specimens (such as the compact specimen) for commonly used engineering materials and thicknesses.

The experimental fracture data for the round robin were gathered by the NASA Langley Research Center and Westinghouse Research and Development Laboratory. Tests were conducted on compact specimens to obtain load against physical crack extension data and failure loads. The NASA Langley Research Center also conducted fracture tests on additional compact specimens (not part of the baseline data supplied to the round-robin participants), middle-crack tension (MT) specimens (formerly center-crack tension specimens), and "structurally configured" specimens (with three circular holes and a crack emanating from one of the holes) subjected to tensile loading. The three-hole-crack tension (THT) specimen simulates the stress-intensity factor solution for a cracked stiffened panel (see Appendix I). The specimen configurations tested are shown in Fig. 1. In addition, tension specimens were also tested to obtain uniaxial stress-strain curves. The three materials tested were 7075-T651 aluminum alloy, 2024-T351 aluminum alloy, and 304 stainless steel.

Eighteen participants from two countries were involved in the predictive round robin. The participants are listed in Table 1. The participants could use any fracture analysis method or methods to predict failure (maximum load) of the compact specimens, the MT specimens, and the THT specimens from the results

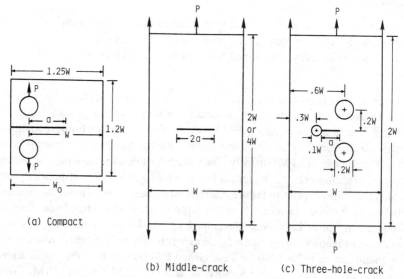

FIG. 1—*Specimen configurations tested and analyzed.*

TABLE 1—*Round robin participants listed in alphabetical order.*

Name	Affiliation
R. J. Allen	British Railways Board
J. M. Bloom	Babcock and Wilcox Co.
G. E. Bockrath	California State University
R. deWit	National Bureau of Standards
J. B. Glassco	Rockwell International Corp.
T. M. Hsu	Gulf Oil E and P Co.
C. M. Hudson	NASA Langley Research Center
J. D. Landes	American Welding Institute
P. E. Lewis	NASA Langley Research Center
B. D. Macdonald	Knolls-Atomic Power Lab.
D. E. McCabe	Westinghouse Electric Co.
P. O. Metz	Armco
J. C. Newman, Jr.[a]	NASA Langley Research Center
D. O'Neal	McDonnell-Douglas Co.
T. W. Orange	NASA Lewis Research Center
D. P. Peng	Monsanto Co.
G. A. Vroman	Rockwell International Corp.
F. J. Witt	Westinghouse Electric Co.

[a]Chairman, ASTM Task Group E24.06.02 on Application of Fracture Analysis Methods.

of tensile and baseline compact specimen fracture data on the three materials. Thirteen different fracture analysis methods were used by the participants. (Only one participant, the author, who submitted two sets of predictions using two different methods, knew the failure loads on all specimens.)

The fracture analysis methods used in the round robin included: linear-elastic fracture mechanics (LEFM) corrected for size effects or for plastic yielding, Equivalent Energy, the Two-Parameter Fracture Criterion (TPFC), the Deformation Plasticity Failure Assessment Diagram (DPFAD), the Theory of Ductile Fracture, the K_R-curve with the Dugdale model, an effective K_R-curve derived from residual strength data, the effective K_R-curve, the effective K_R-curve with a limit-load condition, limit-load analyses, a two-dimensional finite-element analysis using a critical crack-tip-opening displacement criterion with stable crack growth, and a three-dimensional finite-element analysis using a critical crack-front singularity parameter with a stationary crack. Descriptions of these methods, by the participants, are given in the appendices. (These descriptions were written after the results of the round robin were made public.) Table 2 lists the methods used by each participant for each material. Most participants used the same method for all materials, but some participants used different methods for different materials.

The results of the experimental and predictive round robin are discussed in this paper. Comparisons are made between experimental and predictive failure loads on the three specimen types for the three materials. The accuracy of the various methods was judged by the variations in the ratio of predicted-to-experimental failure loads; and the methods were ranked in order of minimum standard

TABLE 2—*Methods used in ASTM E24.06.02 predictive round robin.*

Participant	Material		
	7075-T651	2024-T351	304
1	LEFM (size effects)	LEFM (size effects)	LEFM (size effects)
2	LEFM (plastic zone)	LEFM (plastic zone)	LEFM (plastic zone)
3	Equivalent Energy	Equivalent Energy	Equivalent Energy
4	TPFC (one parameter)	TPFC (one parameter)	TPFC (one parameter)
5	TPFC	TPFC	TPFC
6	DPFAD	DPFAD	DPFAD
7[a]	Theory of Ductile Fracture	Theory of Ductile Fracture	Theory of Ductile Fracture
8[a]	Theory of Ductile Fracture	Theory of Ductile Fracture	Theory of Ductile Fracture
9[a]	Theory of Ductile Fracture	Theory of Ductile Fracture	Theory of Ductile Fracture
10	K_R-curve and Dugdale model	K_R-curve and Dugdale model	...
11	K_R-curve (estimated)	K_R-curve (estimated)	K_R-curve (estimated)
12	K_R-curve	K_R-curve and limit load	...
13	K_R-curve	K_R-curve and limit load	limit load
14	K_R-curve	K_R-curve and limit load	limit load
15	K_R-curve	K_R-curve and limit load	limit load
16	K_R-curve	K_R-curve and limit load	limit load
17	finite-element analysis (2D)	finite-element analysis (2D)	finite-element analysis (2D)
18	finite-element analysis (3D)	finite-element analysis (3D)	finite-element analysis (3D)

[a]Participant used LEFM corrected for size effects for predictions made on compact specimens.

error. The range of applicability of the various methods was also considered in assessing their usefulness. The interpretation of the results in this report is that of the author and may not necessarily be in agreement with the opinions of the participants.

Experimental and Predictive Round Robin

To assess the accuracy and usefulness of many of the elastic-plastic fracture analysis methods, an experimental and predictive round robin was conducted by ASTM Task Group E24.06.02. The objective of the round robin was to determine whether the fracture analysis methods currently used can or cannot predict failure (maximum or instability load) on compact, middle-crack tension, and three-hole-crack tension specimens from the results of tension tests and of compact specimen fracture tests. A brief outline of the experimental and predictive round robin procedure follows.

Materials

7075-T651 aluminum alloy	$B = 12.7$ mm
2024-T351 aluminum alloy	$B = 12.7$ mm
304 stainless steel	$B = 12.7$ mm

Data provided

A. Tensile Properties
   1. Yield stress (0.2% offset)
   2. Ultimate tensile strength
   3. Elastic modulus
   4. Full stress-strain curve
B. Fracture Results on Compact Specimens ($W = 51$, 102, and 203 mm; $a_0/W = 0.5$)
   1. Maximum Failure Loads
   2. Typical load-displacement records
   3. K_R-curve (physical and effective)
   4. J_R-curve
C. Compact, Middle-Crack Tension (MT), and Three-Hole-Crack Tension (THT) Specimens (see Fig. 1)
   1. All specimen dimensions
   2. Initial crack lengths (three-point weighted average through the thickness)
   3. Stress-intensity factor solution for the THT specimen (see Appendix I)

Information Required

Predict the maximum failure load on compact, MT, and THT specimens as a function of initial crack length for the three materials using the data provided.

Experimental Procedure

The experimental test program was conducted by NASA Langley Research Center and Westinghouse Research and Development Laboratory. Tests were conducted on compact specimens (with initial crack-length-to-width ratios, a_0/W, of 0.5) to obtain load against physical crack extension data and failure loads. NASA Langley also conducted fracture tests on other compact specimens (with a_0/W equal to 0.3 and 0.7), MT specimens, and THT specimens. The specimen configurations are shown in Fig. 1. In addition, tension specimens were also tested to obtain uniaxial stress-strain curves.

Materials

The three materials tested were 7075-T651 aluminum alloy, 2024-T351 aluminum alloy, and 304 stainless steel. These materials were selected because they exhibit a wide range in fracture toughness behavior. They were obtained in plate form (1.2 m by 3.6 m) with a nominal thickness of 12.7 mm.

Specimen Configurations and Loadings

Four types of specimens were machined from one plate of each material. The specimens were: (1) tension, (2) compact, (3) middle-crack tension, and (4) three-hole-crack tension specimens. A summary of specimen types, nominal widths, and nominal crack-length-to-width ratios tested is given in Table 3.

Tension Specimens—Eight tension specimens [ASTM Tension Testing of Metallic Materials (E 8-82)] with square cross section (12.7 by 12.7 mm) were machined from various locations in each plate of material. The specimens were machined to obtain tensile properties perpendicular to the rolling direction. Full engineering stress-strain curves were obtained from each specimen. The initial load rate was 45 kN/min, but after yielding, the load rate was set at 4.5 kN/min. Average tensile properties (E, σ_{ys}, and σ_u) are given in Table 4.

Compact Specimens—The compact specimen configuration is shown in Fig.

TABLE 3—*Test specimen matrix and number of specimens for 7075-T651, 2024-T351, and 304 stainless steel.*

Specimen Type	Nominal Width, mm	Nominal Crack-Length-to-Width Ratio			
		0.3	0.4	0.5	0.7
Compact	51	2	...	5[a]	2
Compact	102	2	...	5[a]	2
Compact	203	2	...	5[a]	2
Middle crack	127	...	2	...	...
Middle crack	254	...	2	...	...
Three-hole crack	254	$8(0.05 \leq a_0/W \leq 0.4)$			
Tensile[a]	12.7	...			

[a]Data provided to participants.

TABLE 4—*Average tensile properties of the three materials*[a]

Material	E, MN/m^2	σ_{ys}, MN/m^2	σ_u, MN/m^2	κ, MN/m^2	n
7075-T651	71 700	530	585	640	30
2024-T351	71 400	315	460	550	10
304 stainless steel	203 000	265	630	745	5

[a]Average values for eight tests.

1*a*. The planar configuration is identical to the "standard" compact (ASTM E 399) specimen, but the nominal thickness was 12.7 mm. Twenty-seven specimens were machined from each plate of material, and the cracks were oriented in the same direction (parallel to the rolling direction). The nominal widths, W, were 51, 102, and 203 mm, and the nominal crack-length-to-width ratios were 0.3, 0.5, and 0.7. All specimens were fatigue precracked according to the ASTM E 399 requirements.

The specimens tested by Westinghouse ($a_0/W = 0.5$) were loaded under displacement-control conditions and periodically unloaded (about 15% at various load levels) to determine crack lengths from compliance [5,6]. However, the specimens tested by NASA Langley were loaded under load-control conditions to failure. The initial load rates on the NASA Langley tests were about the same as those tested by Westinghouse. Load against crack extension data were obtained from visual observations and from unloading compliance data (at both the crack mouth and the load line). Initial crack lengths, a_0, and failure loads, P_f, were also recorded. The initial crack lengths were measured from broken specimens and were three-point weighted averages through the thickness ($4a_0 = a_1 + 2a_2 + a_3$) where a_1 and a_3 were surface values and a_2 was the value in the middle of the specimen.

Middle-Crack and Three-Hole-Crack Tension Specimens—The middle-crack and three-hole-crack tension specimen configurations are shown in Figs. 1*b* and 1*c*, respectively. Again, all specimens were machined so that the cracks were oriented parallel to the rolling direction. Four MT specimens ($W = 127$ and 254 mm) were machined from each plate of material. The nominal crack-length-to-width ratio was 0.4. Eight THT specimens ($W = 254$ mm) were also machined from each plate of material. The nominal crack lengths in the three-hole-crack specimen ranged from 13 to 102 mm. All MT and THT specimens were 510 mm between griplines. The initial stress-intensity factor rate was roughly the same (30 MN/m$^{3/2}$/min) for all crack specimens. Again, initial crack lengths (three-point weighted average through the thickness) and failure loads were recorded.

Testing Machines

A 220- and a 1350-kN analog closed-loop servo-controlled testing machines were used to conduct the fracture tests. Figure 2 shows the large test machine with a THT specimen. These systems were used for fatigue precracking and for

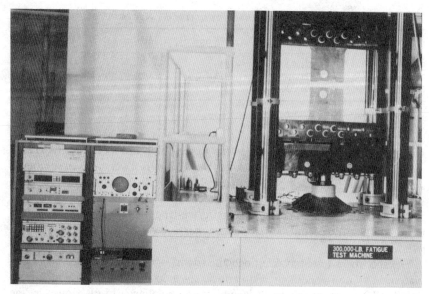

FIG. 2—*Large load capacity fatigue and fracture test machine.*

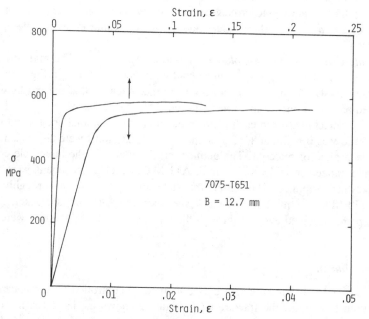

FIG. 3—*Stress-strain curve for 7075-T651 aluminum alloy.*

fracture testing. During the fracture tests, loads were monitored and recorded on an *X-Y* plotter to determine the load at failure.

The test procedures for the compact specimen tests conducted at Westinghouse Research and Development Laboratory [5] are given in Appendix II.

Experimental Results

The following section describes the experimental results obtained from testing tension, compact, middle-crack tension, and three-hole-crack tension specimens. The compact specimens ($a_0/W = 0.5$) tested at Westinghouse Research and Development Laboratory [5] were used to determine effective and physical crack lengths as a function of load. These data were used to develop crack-growth resistance curves in terms of K_R and J_R. The test procedures and typical load-displacement data for these specimens are discussed in Appendix II. Full stress-strain curves, K_R (effective and physical) data, J_R data, and maximum failure loads are presented herein for the three materials.

Aluminum Alloy 7075-T651

Tension Specimens—A typical full engineering stress-strain curve for 7075-T651 aluminum alloy is shown in Fig. 3. The average values of yield stress, ultimate tensile strength, and Young's modulus for eight tests are given in Table 4. The average stress-strain curves were approximated by the Ramberg-Osgood equation [7] as

$$\epsilon = \frac{\sigma}{E} + \left(\frac{\sigma}{\kappa}\right)^n \qquad (1)$$

where κ and n are the strain-hardening coefficient and power, respectively. Values of these constants, fitted to the engineering stress-strain curve, are given in Table 4.

Compact Specimens—A photograph of a large compact fracture specimen ($W = 203$ mm) is shown in Fig. 4a. The 7075-T651 specimen exhibited a very

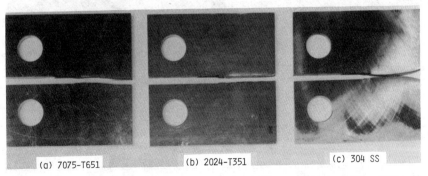

(a) 7075-T651   (b) 2024-T351   (c) 304 SS

FIG. 4—*Photographs of large compact fracture specimens* (W = 203 mm; a_0/W = 0.5) *made of the three materials.*

flat fracture surface appearance, typical of brittle materials. Photographs of the fatigue precrack and fracture surfaces are shown in Fig. 5 for the small compact specimens ($W = 51$ mm) with $a_0/W = 0.3, 0.5$, and 0.7. Note that the fatigue-crack front shape for 7075-T651 specimens is not typical of the shapes commonly observed for fatigue-crack fronts; that is, the crack front in the center is lagging behind other points along the front. Normally, fatigue-crack fronts show the classical "thumbnail" shape.

The effective K_R data for 7075-T651 is shown in Fig. 6. The effective crack extension, Δa_e, was obtained from compliance-indicated crack lengths. Effective crack lengths (a_e) were averages between compliance measurements made at the crack mouth and load line (see Appendix II). The stress-intensity factor was calculated from

$$K = \frac{P}{B\sqrt{W}} \frac{2 + \lambda}{(1 - \lambda)^{3/2}} (0.886 + 4.64\lambda - 13.32\lambda^2 + 14.72\lambda^3 - 5.6\lambda^4)$$

(2)

where $\lambda = a_e/W$ [8]. The symbols show results from the three specimen sizes and show that the K_R data are independent of specimen size. Some discrepancy is observed at large values of crack extension for the 51- and 102-mm-wide specimens.

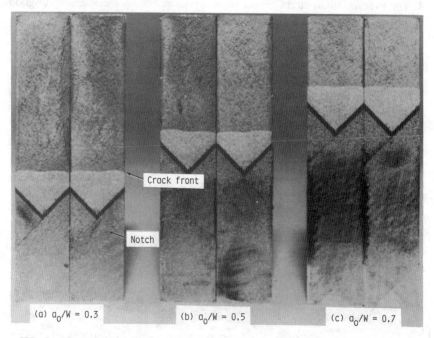

(a) $a_0/W = 0.3$ (b) $a_0/W = 0.5$ (c) $a_0/W = 0.7$

FIG. 5—*Photographs of fatigue-crack growth and fracture surfaces for 7075-T651 aluminum alloy compact specimens (W = 51 mm).*

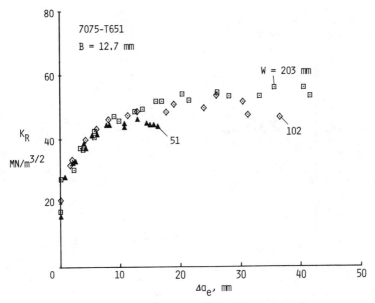

FIG. 6—K_R *against effective crack extension for 7075-T651 aluminum alloy compact specimens* ($a_0W = 0.5$).

The physical K_R data are shown in Fig. 7. The physical crack extension data (Δa_p) were obtained from unloading compliance. Crack lengths determined from compliance were within 5% of visual crack length measurements on the surface. Again, K_R was calculated from Eq 2 using physical crack length (a) instead of a_e. The symbols show that the physical K_R data are independent of specimen size.

The J_R data for 7075-T651 are shown in Fig. 8 for the compact specimens. For this material, J_R was obtained from the physical K_R data by using the elastic relation

$$J_R = K_R^2/E \tag{3}$$

The K_R data and, consequently, the J_R data are independent of specimen size, except for large values of crack extension on each specimen.

Normalized failure loads (P_f/B) on various compact specimens made of 7075-T651 are shown in Fig. 9. The solid symbols show the baseline compact specimen data supplied to the participants. The baseline specimens were tested at Westinghouse and NASA Langley. In general, the average failure loads on the NASA Langley tests (load control) were within ±2% of the average failure loads from the Westinghouse tests (stroke control), except for the 203-mm-wide specimens. Here the Langley test results were about 6% higher than the results from Westinghouse. The open symbols show failure loads on compact specimens with

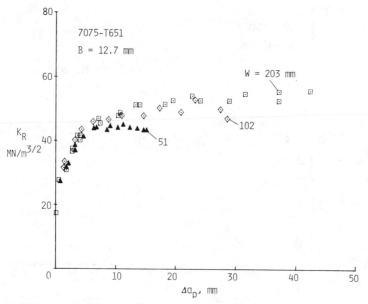

FIG. 7—K_R against physical crack extension from unloading compliance for 7075-T651 aluminum alloy compact specimens ($a_0/W = 0.5$).

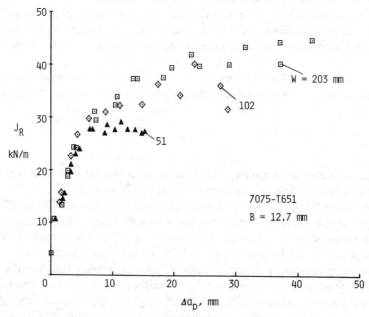

FIG. 8—J_R against physical crack extension for 7075-T651 aluminum alloy compact specimens ($a_0/W = 0.5$).

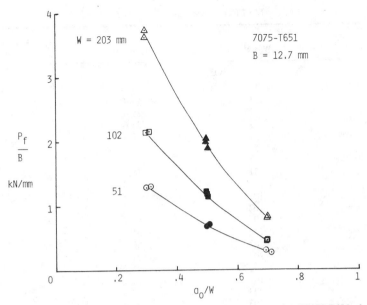

FIG. 9—*Normalized failure loads on various compact specimens made of 7075-T651 aluminum alloy (solid symbols denote data supplied to participants).*

a_0/W = 0.3 and 0.7. These loads were to be predicted by the participants. Table 5*a* gives the failure loads on all 7075-T651 compact specimens.

Middle-Crack Tension Specimens—A photograph of an MT specimen (W = 254 mm) tested at NASA Langley is shown in Fig. 10*a*. Again, the 7075-T651 specimen showed a flat fracture surface appearance indicative of brittle materials.

Figure 11 shows the failure loads on the two size MT specimens tested with a nominal $2a_0/W$ of 0.4. Table 5*b* gives specimen dimensions, average initial crack lengths, and failure loads on MT specimens. These failure loads were to be predicted by the participants.

Three-Hole-Crack Tension Specimens—Figure 12*a* shows a photograph of a THT specimen (a_0 = 25.4 mm) tested by NASA Langley. Motion pictures (200 frames per second) were taken of these specimens. A voltmeter was used to indicate applied load in the movie. Load against crack length measurements taken from these motion pictures are shown in Fig. 13. The initial crack lengths, a_0, were about 25.4 mm. The circle symbols show experimental data on a 7075-T651 aluminum alloy specimen. Solid symbols show the final crack lengths near maximum load conditions. As expected, the final crack lengths were past the centerline of the large holes and were very near the minimum stress-intensity factor location (see Appendix I). A photograph from motion picture frames near maximum (failure) load conditions is shown in Fig. 14*a*.

Failure loads plotted against initial crack length for the THT specimens with W = 254 mm are shown in Fig. 15. For crack lengths less than about 63.5 mm (centerline of large holes), the failure loads (Table 5*c*) were not influenced by

TABLE 5—*Aluminum alloy 7075-T651.*

B, mm	W, mm	a_0, mm	Experimental P_f, kN
	(*a*) Compact Specimens		
12.4	51	16.1	16.1
12.5	51	15.4	16.0
12.7[a]	51	25.6	8.73
12.8[a]	51	25.6	8.85
12.6	51	25.4	8.54
12.6	51	25.9	8.85
12.6	51	35.4	3.75
12.5	51	36.3	3.34
12.7	102	31.8	27.4
12.7	102	30.6	27.2
12.8[a]	102	50.8	15.5
12.7[a]	102	50.7	15.5
12.8[a]	102	51.4	14.5
12.7	102	50.9	15.1
12.8	102	51.4	15.1
12.6	102	71.2	5.78
12.7	102	71.0	5.65
12.7	203	60.6	47.4
12.8	203	60.4	46.3
12.8[a]	203	102.0	24.1
12.7[a]	203	102.2	24.1
12.8[a]	203	100.8	25.4
12.8	203	101.2	25.7
12.8	203	101.2	26.2
12.8	203	142.0	10.2
12.7	203	142.2	10.5
	(*b*) Middle-Crack Tension Specimens		
12.8	127	26.4	209
12.8	127	24.9	200
12.8	254	49.8	365
12.7	254	49.1	356
	(*c*) Three-Hole Crack Tension Specimens		
12.8	254	11.9	696
12.8	254	25.5	685
12.7	254	39.7	698
12.7	254	50.5	651
12.8	254	64.8	620
12.7	254	75.6	578
12.8	254	90.1	462
12.8	254	100.8	362

[a]Tested at Westinghouse Research Laboratory [5].

crack length as much as those for crack lengths greater than 63.5 mm. Again, these failure loads were to be predicted by the participants.

Aluminum Alloy 2024-T351

Tension Specimens—A typical full engineering stress-strain curve for 2024-T351 aluminum alloy is shown in Fig. 16. The yield stress, ultimate tensile

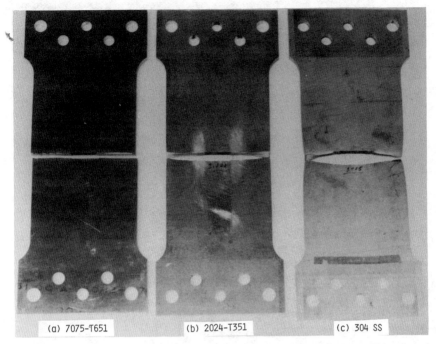

FIG. 10—*Photographs of large middle-crack tension fracture specimens* (W = 254 mm) *made of the three materials.*

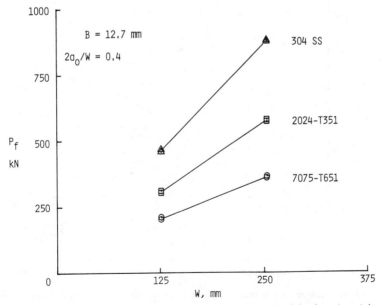

FIG. 11—*Failure loads on middle-crack tension specimens made of the three materials.*

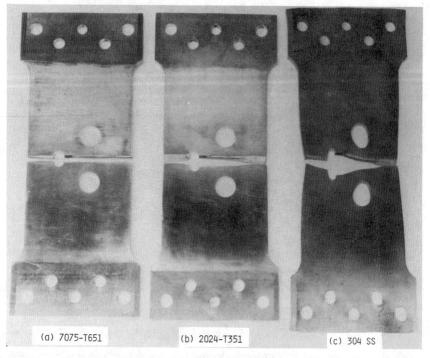

FIG. 12—*Photographs of three-hole-crack tension fracture specimens made of the three materials* (W = 254 mm; a_0 = 25.4 mm).

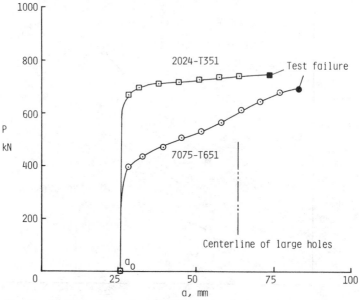

FIG. 13—*Experimental stable crack growth behavior for the three-hole-crack tension specimens made of the aluminum alloys.*

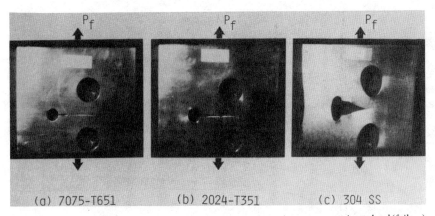

FIG. 14—*Photographs of three-hole-crack tension fracture specimens near maximum load (failure) condition* ($a_0 = 25.4$ mm).

strength, Young's modulus, and Ramberg-Osgood constants are given in Table 4.

Compact Specimens—A photograph of a 2024-T351 aluminum alloy compact specimen is shown in Fig. 4*b*. The fracture surface showed substantial shear lip development during fracture. Figure 17 shows photographs of the fatigue precrack and fracture surfaces for the small compact specimens. The fatigue-crack front

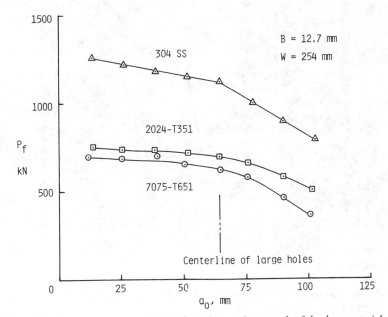

FIG. 15—*Failure loads on three-hole-crack tension specimens made of the three materials.*

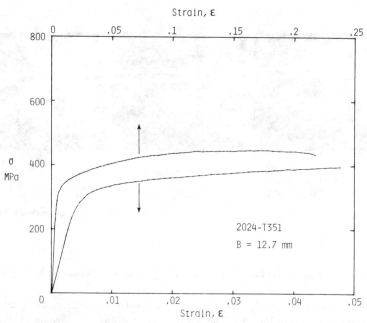

FIG. 16—*Stress-strain curve for 2024-T351 aluminum alloy.*

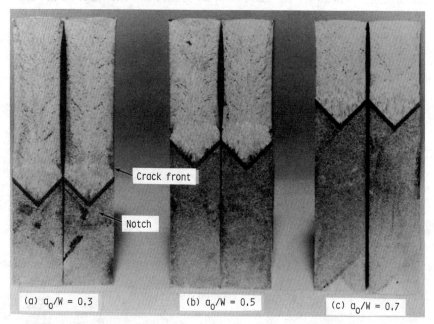

FIG. 17—*Photographs of fatigue-crack growth and fracture surfaces for 2024-T351 aluminum alloy compact specimens (W = 51 mm).*

showed the "thumbnail" shape. The crack length in the center of the specimen was about 1 mm longer than lengths measured on the surfaces.

Effective K_R data for 2024-T351 are shown in Fig. 18. The effective crack extension was, again, obtained from compliance-indicated crack lengths. Equation 2 was used to calculate K_R. The symbols show results from three specimen sizes and show that the K_R-curve is independent of specimen size.

The physical K_R data are shown in Fig. 19. Physical crack extension was, again, obtained from unloading compliance. The physical crack lengths from unloading compliance were within 5% of visual surface measurements. The symbols show the experimental data and show that the physical K_R data are not independent of specimen size near their peak values.

The J_R data plotted against physical crack extension for 2024-T351 are shown in Fig. 20. The J_R values (symbols) were obtained from load-displacement records (V_{LL}) and an equation given by Hutchinson and Paris (Ref 9, Eq 31, p. 47). The equation is

$$J = 2 \int_0^{\theta_c} \frac{M_0}{W - a} d\theta_c - \int_{a_0}^a \frac{J}{W - a} da \qquad (4)$$

where M_0 is the applied moment per unit thickness and θ_c is the rotation due to the presence of the crack. Equation 4 was rewritten as

$$J_R = \frac{A}{B(W - a)} f\left(\frac{a}{W}\right) - \sum J \frac{\Delta a}{W - a} \qquad (5)$$

where A is the area under the load-displacement record and $f(a/W)$ is given by

$$f\left(\frac{a}{W}\right) = 2(1 + \phi)/(1 + \phi^2) \qquad (6)$$

where

$$\phi = \left[\left(\frac{2a_0}{W - a}\right)^2 + 2\left(\frac{2a_0}{W - a}\right) + 2\right]^{1/2} - \left(\frac{2a_0}{W - a} + 1\right) \qquad (7)$$

and the last term in Eq 5 is the summation of $J\Delta a/(W - a)$ from the initial crack length to the specified crack length. Only the results for the 102- and 203-mm-wide specimens were analyzed. The results show that the J_R curve is independent of specimen size.

Normalized failure loads on various-width compact specimens as a function of a_0/W are shown in Fig. 21. Again, the solid symbols show the baseline compact specimen data supplied to the participants. Average failure loads on the NASA Langley tests were within ±2% of the average failure loads from the Westinghouse tests; see Table 6a. The open symbols show results that were to be predicted by the participants.

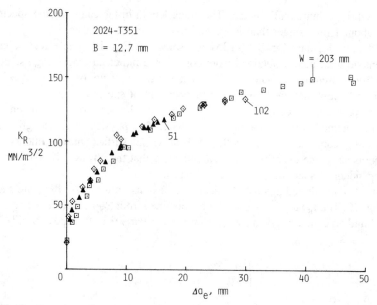

FIG. 18—K_R against effective crack extension for 2024-T351 aluminum alloy compact specimens ($a_0/W = 0.5$).

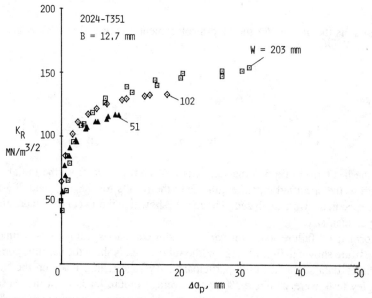

FIG. 19—K_R against physical crack extension from unloading compliance for 2024-T351 aluminum alloy compact specimens ($a_0/W = 0.5$).

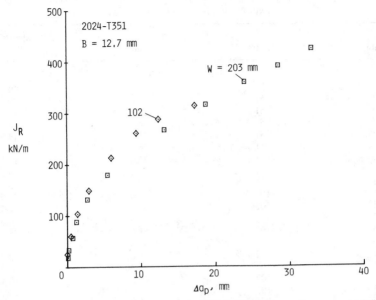

FIG. 20—*J_R against physical crack extension for 2024-T351 aluminum alloy compact specimens* (a₀/W = 0.5).

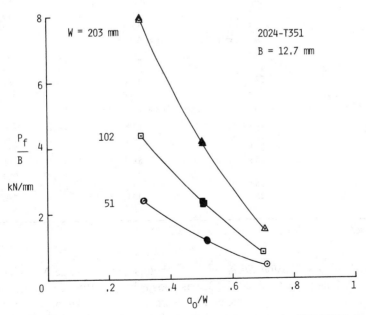

FIG. 21—*Normalized failure loads on various compact specimens made of 2024-T351 aluminum alloy (solid symbols denote data supplied to participants).*

TABLE 6—*Aluminum alloy 2024-T351.*

B, mm	W, mm	a_0, mm	Experimental P_f, kN
		(a) COMPACT SPECIMENS	
12.4	51	16.1	29.8
12.3	51	16.0	29.5
12.6[a]	51	26.5	14.2
12.5[a]	51	26.3	14.7
12.5[a]	51	26.1	14.8
12.3	51	26.1	14.5
12.4	51	26.4	14.7
12.3	51	36.2	5.22
12.3	51	36.3	5.29
12.5	102	31.4	54.7
12.5	102	31.2	54.7
12.5[a]	102	51.9	28.8
12.6[a]	102	51.6	28.9
12.6[a]	102	51.4	29.8
12.5	102	51.6	28.2
12.5	102	51.9	28.7
12.6	102	71.2	10.1
12.5	102	71.4	10.1
12.5	203	61.8	98.5
12.6	203	61.7	100.3
12.6[a]	203	102.4	52.1
12.6[a]	203	102.5	51.9
12.5	203	102.2	52.3
12.6	203	102.2	52.0
12.5	203	142.9	18.6
12.5	203	143.0	18.9
		(b) MIDDLE-CRACK TENSION SPECIMENS	
12.6	127	26.2	302
12.6	127	25.2	311
12.6	254	51.2	581
12.6	254	52.1	574
		(c) THREE-HOLE CRACK TENSION SPECIMENS	
12.6	254	13.9	754
12.5	254	25.7	738
12.5	254	38.6	735
12.5	254	51.8	718
12.6	254	64.3	696
12.6	254	75.8	660
12.5	254	90.0	580
12.5	254	101.5	505

[a]Tested at Westinghouse Research Laboratory [5].

Middle-Crack Tension Specimens—A photograph of a large MT specimen (W = 254 mm) made of 2024-T351 is shown in Fig. 10b. The experimental failure loads on these specimens are shown in Fig. 11. Although the tensile strength of the 2024-T351 material is much lower than that of the 7075-T651 material, the failure loads are much higher for the same initial crack length, width, and thickness. Failure loads and specimen dimensions are given in Table 6*b*.

Three-Hole-Crack Tension Specimens—Photographs of the THT specimens made of 2024-T351 are shown in Figs. 12*b* and 14*b*. Figure 13 shows load against crack length measurements made on the THT specimens with a_0 = 25.4 mm. Although the failure loads and final crack lengths were quite close for the two aluminum alloys, the load-crack-length behavior of the 2024-T351 material was quite different from that of the 7075-T651 material. The failure loads as a function of initial crack length are shown in Fig. 15 (see Table 6*c*). The failure loads on the 2024-T351 specimens were consistently higher than those on the 7075-T651 specimens.

Stainless Steel 304

Tension Specimens—A typical full engineering stress-strain curve for 304 stainless steel is shown in Fig. 22. A summary of the average tensile properties and the Ramberg-Osgood constants is given in Table 4.

Compact Specimens—A photograph of a 304 stainless steel specimen is shown in Fig. 4*c*. The specimen exhibited very large deformations along the crack line during fracture. The thickness of the material along the crack line contracted to about 65% of the original thickness. Photographs of the fatigue precrack and fracture surfaces are shown in Fig. 23. Again, the fatigue-crack front showed

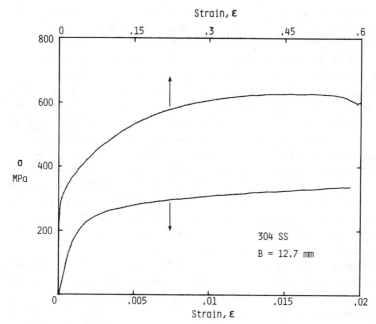

FIG. 22—*Stress-strain curve for 304 stainless steel.*

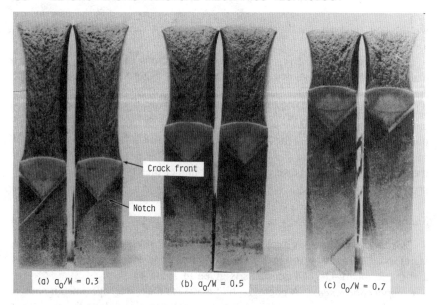

FIG. 23—*Photographs of fatigue-crack growth and fracture surfaces for 304 stainless steel compact specimens* (W = 51 mm).

the classical "thumbnail" shape. The crack length in the center of the specimen was about 1.3 mm longer than lengths measured on the surfaces.

The effective K_R data for 304 stainless steel are shown in Fig. 24. Again, the effective crack extension values were obtained from compliance-indicated crack lengths. The effective K_R data for this material was dependent upon specimen size. Smaller specimen widths gave higher K_R values for a given Δa_e.

Figure 25 shows the physical K_R data for the three specimen sizes tested. Here the physical crack extensions (Δa_p) were obtained from visual observations. A comparison between crack lengths obtained from unloading compliance and those from visual observations was not good. Therefore, the visual crack lengths were used. Here, again, the physical K_R data are specimen size dependent.

J_R data for 304 stainless steel compact specimens ($a_0/W = 0.5$) are shown in Fig. 26. The J_R data were obtained by using the same procedure as described for the 2024-T351 material (see Eqs 4 through 7). For this material, the physical crack extensions were obtained from visual observations. Again, only the 102- and 203-mm-wide specimens were analyzed. These results show that the J_R data are dependent upon specimen size. The data for the 102-mm-wide specimen are higher than that for the 203-mm-wide specimen.

Normalized failure loads on various-width compact specimens as a function of a_0/W are shown in Fig. 27. Solid symbols show the baseline compact specimen data supplied to the participants. The average failure loads on the NASA Langley tests (load control) were, generally, within ±2% of the average failure loads from the Westinghouse tests (stroke control), except for the large specimens

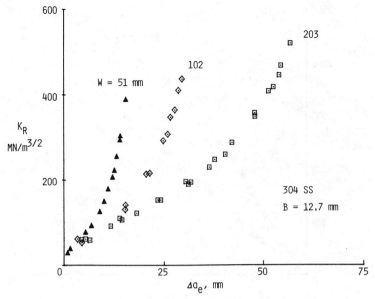

FIG. 24—K_R against effective crack extension for 304 stainless steel compact specimens ($a_0/W = 0.5$).

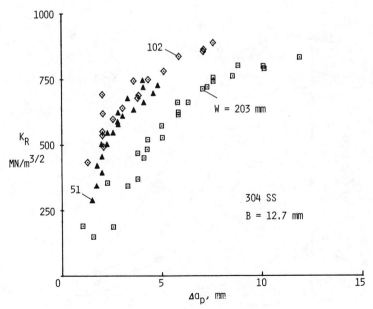

FIG. 25—K_R against physical crack extension from visual crack length measurements for 304 stainless steel compact specimens ($a_0/W = 0.5$).

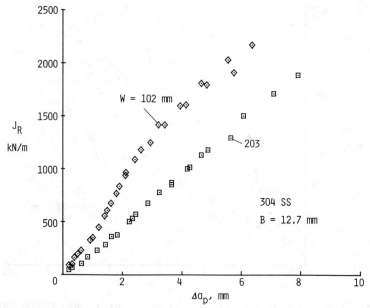

FIG. 26—J_R against physical crack extension (visual measurements) for 304 stainless steel compact specimens ($a_0/W = 0.5$).

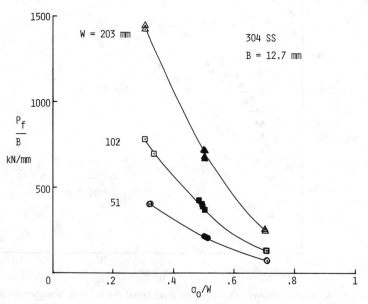

FIG. 27—Normalized failure loads on various compact specimens made of 304 stainless steel (solid symbols denote data supplied to participants).

(W = 203 mm). The failure loads on the Langley tests were about 6% higher than those from the Westinghouse tests (see Table 7a). The open symbols show results that were to be predicted by the participants.

Middle-Crack Tension Specimens—A photograph of one of the 304 stainless steel MT specimens (W = 254 mm) is shown in Fig. 10c. Experimental failure

TABLE 7—*Stainless steel 304.*

B, mm	W, mm	a_0, mm	Experimental P_f, kN
		(a) COMPACT SPECIMENS	
13.1	51	16.5	52.7
13.3	51	16.3	53.6
12.8[a]	51	25.6	27.3
12.8[a]	51	26.1	25.9
12.8[a]	51	25.8	26.8
13.1	51	25.8	27.5
13.1	51	26.2	26.9
13.2	51	36.2	9.56
13.1	51	36.2	9.61
13.4	102	34.1	93.4
13.3	102	31.1	104
13.0[a]	102	49.4	55.1
13.0[a]	102	50.7	50.8
13.0[a]	102	51.4	47.8
13.0	102	50.5	51.8
13.3	102	51.8	50.6
13.4	102	72.1	17.7
13.3	102	72.3	17.3
13.5	203	62.0	195
13.5	203	62.0	192
12.8[a]	203	102.0	86.8
12.8[a]	203	102.3	85.4
12.8[a]	203	102.0	85.3
13.4	203	101.4	96.3
13.4	203	102.2	96.1
13.3	203	142.6	34.1
13.4	203	142.8	32.9
		(b) MIDDLE-CRACK TENSION SPECIMENS	
13.6	127	26.1	458
13.6	127	26.2	469
13.5	254	50.1	882
13.6	254	50.8	878
		(c) THREE-HOLE CRACK TENSION SPECIMENS	
13.6	254	13.5	1260
13.6	254	26.3	1220
13.5	254	39.1	1180
13.4	254	51.6	1150
13.5	254	64.4	1120
13.5	254	77.7	999
13.6	254	89.8	895
13.6	254	102.7	790

[a]Tested at Westinghouse Research Laboratory [5].

loads on these specimens are given in Table 7b and are shown in Fig. 11. The failure loads on the steel specimens were considerably higher than those for the two aluminum alloys.

Three-Hole-Crack Tension Specimens—Photographs of the THT specimens made of 304 stainless steel are shown in Figs. 12c and 14c. These photographs show the large plastic deformations that occur in this material during fracture. Load against crack lengths, such as those shown in Fig. 13, were not recorded on this specimen because the extreme deformation causes uncertainties in measuring physical crack lengths. These crack lengths would have no physical meaning in the undeformed state. The failure loads for various initial crack lengths are shown in Fig. 15. Again, the failure loads for this material were considerably higher than those for the two aluminum alloys. Tabulated failure loads are given in Table 7c.

Predictive Methods Used in Round Robin

The fracture analysis methods used in the round robin were:

1. LEFM corrected for size effects.
2. LEFM corrected for plastic yielding at crack tip.
3. Equivalent Energy.
4. Two-Parameter Fracture Criterion (TPFC).
5. Deformation Plasticity Failure Assessment Diagram (DPFAD).
6. Theory of Ductile Fracture.
7. K_R-curve with Dugdale model.
8. Effective K_R-curve derived from failure load data.
9. Effective K_R-curve.
10. Effective K_R-curve with limit-load criterion.
11. Limit-load (or plastic-collapse) analyses.
12. Two-dimensional finite-element analysis (stable crack growth).
13. Three-dimensional finite-element analysis (stationary crack).

This section briefly describes the methods used by the participants to predict failure loads on compact, middle-crack, and three-hole-crack tension specimens. Table 2 gives a summary of the methods used by the 18 participants for each material. The reader is referred to the Appendices for more details on particular methods.

Participant 1 used a critical elastic stress-intensity-factor approach accounting for the effects of specimen width. Using baseline compact specimen data, a "K_{Ie} against specimen width" curve was constructed for each material. K_{Ie} was calculated from the elastic stress-intensity factor equations for the compact specimen (Eq 2) using the failure loads and initial crack lengths. Predicted failure loads for the other cracked specimens were obtained by using the stress-intensity factor (K) equation for the particular specimen type and a K_{Ie} value interpolated or extrapolated from the "K_{Ie}-curve" at the desired specimen width. Failure was assumed to occur when K at a_0 was equal to K_{Ie} (see Appendix III).

Participant 2 used a critical stress-intensity-factor approach with the Irwin plastic-zone radius added to initial crack length. Using baseline compact specimen data, a "K_c against specimen width" curve was constructed for each material. K_c was calculated from the elastic stress-intensity factor equation (Eq 2) using the failure load and an effective crack length. The effective crack length, a_e, was the initial crack length, a_o, plus Irwin's plane-strain plastic-zone radius, r_p.

Failure loads on compact specimens were predicted from the stress-intensity factor equation written in terms of the effective crack length and a K_c value obtained from the "K_c against specimen width" curve at the desired width. The predicted failure load was given by

$$P = \frac{K_c B \sqrt{W}}{F(a_e/W)} \tag{8}$$

where $F(a_e/W)$ is the boundary-correction factor and

$$a_e = a_0 + 0.056\left(\frac{K_c}{\sigma_{ys}}\right)^2 \tag{9}$$

For the MT and THT specimens, an average value of critical stress-intensity factor, $\bar{K}_c$, was used to predict failure loads. The $\bar{K}_c$ values are given in Table 8. The procedure used to predict failure loads was like that used on the compact specimens except that K_c was replaced by $\bar{K}_c$. (See Appendix IV for more details.)

Participant 3 used the Equivalent Energy method. The basis for this method lies in the uniqueness of the volumetric energy ratio as discussed in Ref 10. From this uniqueness, the failure stress equation is

$$S_d = \frac{K_{Icd}}{\sqrt{\pi a_0}\, F_d} \tag{10}$$

where K_{Icd} is a critical toughness parameter derived from a "standard" fracture toughness specimen (ASTM E 399) with thickness equal to that of the structure of interest. F_d is a correction factor calibrated to the particular stress (S_d), crack

TABLE 8—Plastic-zone corrected fracture toughness values.

Material	$\bar{K}_c$, MN/m$^{3/2}$
7075-T651	36.3
2024-T351	78.5
304	141.5

TABLE 9—*Fracture toughness parameters from TPFC.*

Material	K_F, MN/m$^{3/2}$	m
7075-T651	40.8	0.36
2024-T351	269.5	0.99
304	1365.0	1.0

size (a_0), and specimen type. Appendix V gives further details of the Equivalent Energy method.

Participant 4 used a one-parameter version of the Two-Parameter Fracture Criterion (TPFC) [11,12]. An experimental relationship had previously been observed [11] between the two parameters (K_F and m) for various steels, aluminum alloys, and titanium alloys. This relationship was

$$m = \tanh(21\, K_F/E) \qquad (11)$$

where K_F/E is given in mm$^{1/2}$. The parameter m is nondimensional. Equation 11 was used to eliminate m from the fracture criterion. Thus, only one test was needed, in principle, to evaluate K_F. The value of K_F for each material was determined from the baseline compact specimen data using a least-squares procedure [11]. The value of K_F and the corresponding m value, from Eq 11, are given in Table 9. The procedure used to predict failure loads on compact and MT specimens is given in Ref 12. The procedure used for the THT specimen is given in Appendix VI.

Participant 5 used a modified Two-Parameter Fracture Criterion. The two parameters, K_F and m, were determined from an analysis of the baseline compact specimen data. The fracture parameters determined are given in Table 10. (Note that in Table 10 the parameter m was not truncated to 1.0, as recommended [11], whenever m is greater than unity.)

The predicted failure loads on compact and MT specimens were computed from equations given in Ref 12. The predicted failure loads on the THT specimens were computed by using the MT specimen equations [12] with the total crack length ($2a_0$ in the MT specimen) set equal to the hole diameter ($D = 25.4$ mm) plus the initial crack length, a_0, in the THT specimen. The ultimate plastic-hinge stress, S_u, was also set equal to σ_u, the same as that for the MT specimen.

Participant 6 used the Deformation Plasticity Failure Assessment Diagram (DPFAD) approach. This approach is derived from the Failure Assessment Diagram (FAD). In the United Kingdom, the FAD is referred to as the R-6 diagram. This approach was based on the two-criteria method of Dowling and Townley [13] which states that a cracked structure will fail by either brittle fracture or plastic collapse; and that these two mechanisms are connected by a transition curve based on the strip yield model [14]. Harrison, et al [15] reformulated the two-criteria method into the R-6 diagram. The FAD method is similar to the J$_R$-

TABLE 10—*Fracture toughness parameters from modified TPFC.*

Material	K_F, MN/m$^{3/2}$	m
7075-T651	70.2	1.96
2024-T351	303.9	1.06
304	773.6	0.74

curve method with a limit-load condition. In other words, the FAD is the material response or failure curve.

Because 7075-T651 aluminum alloy had very little strain hardening, the original R-6 failure assessment diagram was used. For 2024-T351 aluminum alloy and 304 stainless steel, however, the failure assessment diagrams were based upon a power-law hardening true stress-true plastic strain curve for the respective materials [16]. The DPFAD approach is more accurate than the original FAD because it accounts for the actual strain hardening properties of the material. The DPFAD curves used were based on deformation plasticity solutions of an infinite middle-crack specimen under plane-stress conditions.

To predict failure load, the coordinate point (K_r, S_r) on the FAD was determined at a fixed load level such that the coordinate point would lie inside the FAD. The expressions used to calculate these coordinate points were

$$K_r^2 = J_{\text{Ie}}(a + \Delta a_p)/J_{\text{R}}(\Delta a_p) \tag{12}$$

and

$$S_r = S/S_L(a + \Delta a_p) \tag{13}$$

where

$J_{\text{Ie}}(a + \Delta a_p)$ = J-integral calculated from the elastic stress-intensity factor for the desired specimen (K^2/E) with a crack of "$a + \Delta a_p$",

J_{R} = experimental crack-growth resistance (J) curve for the particular material and specimen size,

S = applied stress, and

$S_L(a + \Delta a_p)$ = plastic-collapse stress at "$a + \Delta a_p$."

Each increment of crack growth (Δa_p) produces a point on the FAD. Maximum (failure) load occurs when the coordinate point, when properly scaled, is just outside of the FAD. (See Appendix VIII for details.)

The J_{R}-curves used in Eq 12 were obtained from the baseline compact specimen data and were, in general, a function of specimen size. See Appendix VIII for the particular J_{R}-curves used for the compact, MT, and THT specimens.

Participants 7, 8, and 9 used the Theory of Ductile Fracture [17] to predict failure loads on MT and THT specimens. However, they used LEFM analyses corrected for specimen width effects, like Participant 1, to predict failure on the compact specimens. This brief description will concentrate on the Theory of Ductile Fracture.

Basically, the theory predicts failure of cracked tensile-loaded specimens from information derived from full stress-strain curves of the material of interest and from thickness related parameters (predetermined from a large amount of fracture data in the literature). The theory is based on the observation that fracture stress plotted against crack length for a large width panel is a straight line on a log-log plot. The influence of specimen width on fracture has also been included. Two fracture parameters were used to describe the straight line. One was the slope of the straight line, ω, and the other was crack length, a_u, at the ultimate tensile strength [17]. These parameters were subsequently written in terms of properties from full stress-strain curves and other experimentally determined thickness-related factors. The properties from the stress-strain curve were true ultimate stress, true fracture stress, plastic necking strain, engineering fracture strain, true fracture strain, true plastic strain at fracture, true ultimate strain, true ultimate plastic strain, plastic strain-energy density, true yield stress, true plastic yield strain, exponent in Ramberg-Osgood [7] stress-strain equation, true elastic-limit stress, and true elastic-limit strain. The thickness-related properties were maximum thickness for plane stress, t_o, and three other parameters (k, μ, and β [17]). The failure stress equation is written in terms of all of these properties. Further details are given in Appendix IX and Ref 17.

Participant 10 used the K_R-curve with the Dugdale model. The expression that was fitted to the K_R against physical crack extension (Δa_p) data was

$$K_R^2 = \frac{c_1^2 \, \Delta a_p}{c_2 + \Delta a_p} \tag{14}$$

where c_1 and c_2 were functions of specimen width and material. See Appendix X for the values of c_1 and c_2 used for 7075-T651 and 2024-T351 aluminum alloy specimens. (The 304 stainless steel specimens were not considered by this participant.)

The crack-driving-force curve was based on the Dugdale model [14] and extended to finite configurations by Heald et al [18]. The crack-driving force is given by the following expression for the effective stress-intensity factor

$$K_e = \sigma_o FN \left[\frac{8a}{\pi} \, \ell n \, \sec \left(\frac{\pi S}{2N\sigma_o} \right) \right]^{1/2} \tag{15}$$

where

F = boundary-correction factor,
N = conversion factor on nominal stress,
S = applied stress, and
σ_o = effective flow strength.

From the R-curve concept, the failure load is determined by finding the value of load (or applied stress) at which the crack-driving force curve (Eq 15) becomes tangent to the K_R-curve (Eq 14). Further details are given in Appendix X.

Participant 11 used an estimated effective K_R-curve for each material. In contrast to the usual effective K_R-curve, the K_R-curve was estimated by using only the residual strength data provided on the baseline compact specimens. Following the procedure given in Ref *19*, each fracture test gives one point on the estimated K_R-curve. So that instability predictions could be done numerically, simple equations were fitted to each set of points obtained for each material. These equations were

MATERIAL $\qquad\qquad$ K_R CURVE EQUATION

$$7075\text{-}T651 \qquad K_R^2 = 3388\ (1\ -\ e^{-0.1378\Delta a_e})\quad \text{for}\quad \Delta a_e < 15\ \text{mm} \qquad (16)$$

$$2024\text{-}T351 \qquad K_R^2 = 1330\Delta a_e^{0.8738}\quad \text{for}\quad \Delta a_e < 18\ \text{mm} \qquad (17)$$

$$304 \qquad K_R^2 = 4396\Delta a_e^{0.8026}\quad \text{for}\quad \Delta a_e < 15\ \text{mm} \qquad (18)$$

where Δa_e is in millimetres and K_R is in MN/m$^{3/2}$. For the aluminum alloys, these equations agreed well with the measured effective K_R-curves. However, for the stainless steel, the estimated and measured curves had similar magnitudes but considerably different shapes.

Because of limits on Δa_e and on size requirements (see Appendix XI), predicted failure loads were not reported for MT and THT specimens made of 2024-T351 aluminum alloy and 304 stainless steel. For these materials, it was assumed that a limit-load failure would occur, but no attempt was made to calculate limit loads. Further details are given in Appendix XI.

Participant 12 used the effective K_R-curve method for the 7075-T651 aluminum alloy and a net-section stress (limit-load) criterion for 2024-T351 aluminum alloy (see Appendix XII for details). Predictions were not made on 304 stainless steel specimens. He made predictions on only the THT specimens.

The effective K_R-curve, experimentally determined from the baseline compact specimens, was used to predict failure loads on the 7075-T651 THT specimens. The procedure used was identical to the K_R-curve practice described in the ASTM Recommended Practice for R-Curve Determination (E 561-81). Crack-driving force curves for the THT specimen were calculated from the elastic stress-intensity factor equation (see Appendix I). Failure load was then determined by finding the value of load such that the crack-driving force curve would be tangent to the effective K_R-curve.

Participants 13–16 used basically the same fracture-analysis methods. They used the effective K_R-curve for 7075-T651 aluminum alloy specimens, the effective K_R-curve with a limit-load criterion for 2024-T351 aluminum alloy specimens, and limit-load (or plastic-collapse) criteria for 304 stainless steel specimens. Participant 15, however, did deviate from the usual K_R-curve practice.

For 7075-T651 aluminum alloy, the effective K_R-curve experimentally deter-

TABLE 11—*Critical CTOD values from finite-element analysis.*

Material	δ_c, mm
7075-T651	0.0216
2024-T351	0.0457
304	0.357

mined from the baseline compact specimens was used to predict failure loads on all cracked specimens. The procedure used was identical to the K_R-curve practice described in ASTM E 561. That is, the crack-driving force for the three specimen types was calculated from the elastic stress-intensity factor equations. The failure load was then determined by finding the value of load at which the crack-driving force curve became tangent to the effective K_R-curve.

For 2024-T351 aluminum alloy, the effective K_R-curve concept with a limit-load criterion was used to predict failure loads on all cracked specimens. Again, the crack-driving force curve for the three specimen types were calculated from elastic stress-intensity factor equations. The failure load was determined by finding the value of load at which the crack-driving force curve became tangent to the effective K_R-curve, if the load was less than a specified limit load. Otherwise, the failure load was predicted by the limit load. The particular values of limit load used by the various participants were different and they are discussed in Appendices XIII, XIV, and XV.

For 304 stainless steel specimens, limit-load analyses were used to predict failure loads on all specimens. The particular limit-load values or equations used by the various participants are given in Appendices XIII–XV.

Participant 17 used a two-dimensional, elastic-plastic, finite-element analysis which included the effects of stable crack growth [20]. An incremental and small strain finite-element analysis under plane-stress conditions, in conjunction with a crack-tip-opening-displacement (CTOD) fracture criterion, was used to model the crack-growth behavior (initiation, stable crack growth and instability) under monotonic loading to failure. A critical value of CTOD (δ_c) at a specified distance (d) from the crack tip was the fracture criterion. [This is also equivalent to a critical crack-tip-opening-angle (CTOA) criterion.]

The critical value of CTOD was determined from the experimental load against physical crack extension data on the baseline compact specimens. The critical CTOD values determined from the finite-element analysis are given in Table 11. The distance d (or mesh size) along the crack line for all three materials and all specimen sizes was chosen as 0.4 mm.

The failure loads on the compact, MT, and THT specimens were predicted using the finite-element analysis with the δ_c values ($d = 0.4$ mm) determined from the baseline compact specimens. Predictions were not made on some of the 304 stainless steel specimens because of the high computer cost involved. Further details are given in Appendix XVI.

Participant 18 used a three-dimensional, elastic-plastic, finite-element analysis

TABLE 12—*Critical crack-front singularity parameters.*

Material	K_f, MN/m$^{3/2}$
7075-T651	32.8
2024-T351	40.8
304	99.0

with stationary cracks [21]. The stress-strain curves were represented as multilinear strain-hardening behavior. The crack-tip singularity for a multilinear strain-hardening material is $r^{-1/2}$, the same as for a linear-elastic material. However, the strength of the elastic-plastic stress singularity, K_f, is not equal to the elastic stress-intensity factor, K.

The finite-element analysis was used to determine the fracture parameter, K_f, from the baseline compact specimen ($W = 203$ mm) data. Using initial crack length, the applied load on the finite-element model of the specimen was increased incrementally until the experimental failure load was reached. At this load, K_f was computed as a through-the-thickness weighted average. The values computed for each material are given in Table 12.

The predicted failure loads on only a few of the compact and three-hole-crack tension specimens were made using the finite-element analysis. Using initial crack length, the failure load was predicted by incrementally loading the finite-element model of the desired specimen configuration until the critical value of K_f was reached. Further details are given in Appendix XVII.

Predicted Results

Eighteen participants used the fracture analysis methods outlined in Table 2, and discussed in detail in the appendices, to predict failure of compact, middle-crack, and three-hole-crack tension specimens. They used fracture data from various-size compact specimens ($a_0/W = 0.5$) and tensile properties to obtain various fracture parameters for the three materials. The accuracy of the prediction methods was judged by the variations in the ratio of predicted-to-experimental failure loads. Comparisons of the range and mean of the ratio of predicted-to-experimental failure loads are made among the various prediction methods for each specimen type. For the three-hole-crack specimens, some selected methods are compared on how well they can predict failure load as a function of initial crack length.

For each prediction method, a standard error was computed for each specimen type as

$$\text{SE} = \sqrt{\frac{\sum_{i=1}^{M}(1 - P_p/P_f)_i^2}{M}} \qquad (19)$$

where

M = number of predictions submitted by the participant,
P_p = predicted failure load, and
P_f = experimental failure load.

The prediction methods were ranked in order of minimum standard error. The range of applicability of the prediction methods was also considered in assessing their usefulness.

Aluminum Alloy 7075-T651

Compact Specimens—The fracture data on compact specimens with a nominal crack-length-to-width ratio of 0.5 were used by most participants to obtain their fracture constant or parameters. Three participants (7, 8, and 9) determined their fracture parameters from tensile stress-strain data only. They used a different method to predict failure on compact specimens than that used on MT and THT specimens.

Each participant was asked to calculate failure loads on the baseline compact specimens to see how well his method could correlate the baseline data. Figure 28 shows the range and mean of the ratio of calculated-to-experimental failure loads on the 7075-T651 baseline data for each participant. Most participants, who submitted calculations, correlated the baseline data quite well (within $\pm 10\%$).

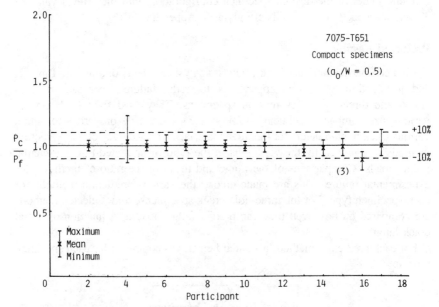

FIG. 28—*Range of ratio of calculated to experimental failure loads on 7075-T651 aluminum alloy compact specimen data supplied to participants (14 tests; number in parentheses denotes calculations submitted).*

Participant 15 submitted only three typical calculations. The one-parameter version of the TPFC (Participant 4) and one application of the effective K_R-curve (Participant 16) had by far the largest errors in failure load calculations (see Table 13).

Compact specimens with $a_0/W = 0.3$ and 0.7 were also tested to see if the methods used could predict the effects of a_0/W on failure loads. These results were unknown to all participants except Participants 4 and 17. Figure 29 shows the range and mean of the ratio of predicted-to-experimental failure loads on these compact specimens. Only about one half of the participants could predict failure loads within $\pm 10\%$ of experimental loads. Surprisingly, the LEFM methods (Participants 1, 2, and 7–9) and a method which reduces to LEFM for brittle materials (Participant 4) had difficulty in predicting failure loads on this material. A summary of standard errors on the compact specimens is given in Table 13.

Middle-Crack Tension Specimens—A comparison of predictions made on MT specimens is shown in Fig. 30. Most participants underestimated the failure loads by about 10% except Participants 7–9. They greatly overestimated the failure loads (see Appendix IX for an explanation). The extremely low predictions from Participant 16 were due to using the wrong equation for the MT specimen. Table 13 gives a summary of standard errors as these specimens.

Three-Hole-Crack Tension Specimens—The range and mean of the ratio of predicted-to-experimental failure loads for the THT specimens are shown in Fig. 31. Again, most participants underestimated the failure loads, except Participants

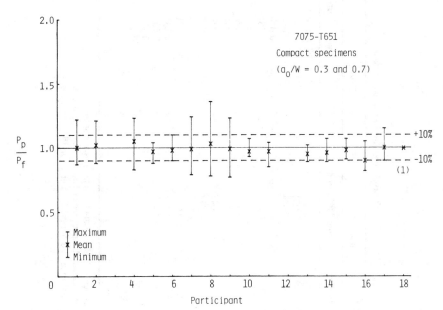

FIG. 29—*Range of ratio of predicted to experimental failure loads on 7075-T651 aluminum alloy compact specimens (twelve tests; number in parentheses denotes predictions submitted).*

TABLE 13—Ranking of participants in order of minimum standard error on 7075-T651 aluminum alloy.

Compact[a]		Compact		MT		THT	
Participant	Standard Error	Participant	Standard Error	Participant	Standard Error	Participant	Standard Error
7	0.023	18	0.004[b]	17	0.073	11	0.072[b]
5	0.024	10	0.049	11	0.078	15	0.095
2	0.024	15	0.054	1	0.084	14	0.097
9	0.024	5	0.058	15	0.115	12	0.113
10	0.025	13	0.061	14	0.140	17	0.118
11	0.032	11	0.065	13	0.141	13	0.121
8	0.032	6	0.070	10	0.152	16	0.133
6	0.035	17	0.075	6	0.159	10	0.149
14	0.042	14	0.076	4	0.181	6	0.216
13	0.047	1	0.082	2	0.195	18	0.278[b]
15	0.055[b]	2	0.104	5	0.227	1	0.331
17	0.063	16	0.121	3	0.329	4	0.354
16	0.119	4	0.141	16[b]	0.512	2	0.365
4	0.121	9	0.155	9	0.587	3	0.442
1	...[c]	7	0.157	7	0.626	5	0.455
3	...[c]	8	0.229	8	0.633	7	1.05
12	...[c]	3	...[c]	12	...[c]	8	1.05
18	...[c]	12	...[c]	18	...[c]	9	1.05

[a]Baseline fracture data provided to participants.
[b]Participant did not predict failure on all specimens.
[c]Participant did not predict failure on these specimens.

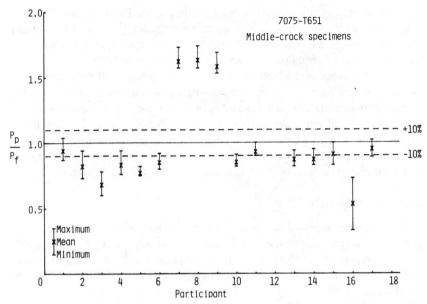

FIG. 30—*Range of ratio of predicted to experimental failure loads on 7075-T651 aluminum alloy middle-crack specimens (four tests).*

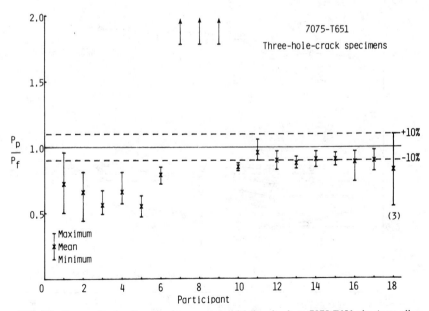

FIG. 31—*Range of ratio of predicted to experimental failure loads on 7075-T651 aluminum alloy three-hole-crack specimens (eight tests; number in parentheses denotes predictions submitted).*

7–9. Again, they greatly overestimated the failure loads. Participants 1–5 greatly underestimated the failure loads. The K_R-curve methods (Participants 10–16) and the finite-element analysis using a CTOD criterion (Participants 17) were the best. The standard errors on these specimens are given in Table 13.

In Fig. 32, some of the methods are compared on how well they can predict failure load as a function of initial crack length. Participants 11 and 15 had the best predictions and the lowest standard errors (see Table 13) for these specimens. Participant 11 did not make predictions for crack-length-to-width ratios lower than 0.25 because of limits on the amount of crack extension allowed with his method (see Appendix XI). With the exception of Participant 1, most participants had about the same ratio of P_p/P_f as a function of initial crack length.

Aluminum Alloy 2024-T351

Compact Specimens—The range and mean of the ratio of calculated-to-experimental failure loads on the baseline compact specimen data are shown in Fig. 33. Remarkably, all methods were able to correlate the fracture data extremely well on this material. Most failure load calculations were within ±5% of the experimental loads.

Many of the methods were also able to predict failure loads on the other compact specimens (a_0/W = 0.3 and 0.7) within ±10% of experimental failure loads, as shown in Fig. 34. Only the LEFM methods, 1 and 2, had some difficulty. A summary of the standard errors on the compact specimens is given in Table 14.

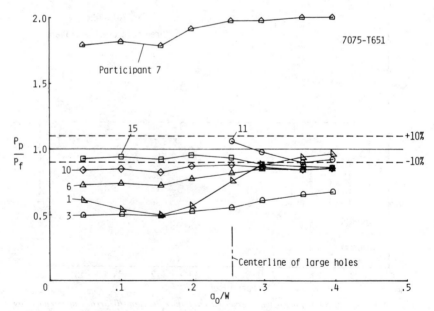

FIG. 32—*Ratio of predicted to experimental failure load as a function of crack length for 7075-T651 three-hole-crack specimens for some selected participants.*

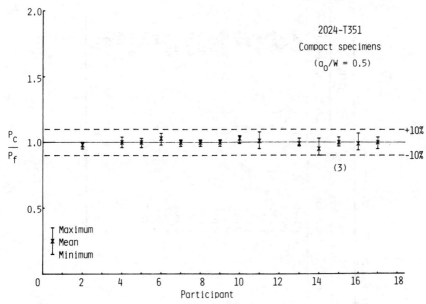

FIG. 33—*Range of ratio of calculated to experimental failure loads on 2024-T351 aluminum alloy compact specimen data supplied to participants (14 tests; number in parentheses denotes calculations submitted).*

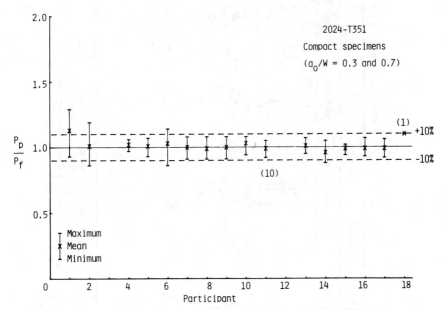

FIG. 34—*Range of ratio of predicted to experimental failure loads on 2024-T351 aluminum alloy compact specimens (twelve tests; number in parentheses denotes predictions submitted).*

TABLE 14—Ranking of participants in order of minimum standard error on 2024-T351 aluminum alloy.

Compact[a]		Compact		MT		THT	
Participant	Standard Error	Participant	Standard Error	Participant	Standard Error	Participant	Standard Error
7	0.013	15	0.027	5	0.035	17	0.040
9	0.013	13	0.038	10	0.036	4	0.041
13	0.016	4	0.039	4	0.050	6	0.058
8	0.017	5	0.041	17	0.051	5	0.087
5	0.020	11	0.046[b]	13	0.057	3	0.087
4	0.020	17	0.049	9	0.080	10	0.102
2	0.023	10	0.051	8	0.082	14	0.122
15	0.026[b]	16	0.055	7	0.090	18	0.133[b]
17	0.030	7	0.057	14	0.113	2	0.235
10	0.031	9	0.059	6	0.144	16	0.251
6	0.038	8	0.060	2	0.149	15	0.333
16	0.042	14	0.064	3	0.190	8	0.346
11	0.044	6	0.098	15	0.234	9	0.348
14	0.067	18	0.101[b]	1	0.401	7	0.348
1	...[c]	2	0.120	16	0.470	12	0.364
3	...[c]	1	0.179	11	...[c]	13	0.476
12	...[c]	3	...[c]	12	...[c]	1	0.559
18	...[c]	12	...[c]	18	...[c]	11	...[c]

[a]Baseline fracture data provided to participants.
[b]Participant did not predict failure on all specimens.
[c]Participant did not predict failure on these specimens.

Middle-Crack-Tension Specimens—The range and mean of the ratio of predicted-to-experimental failure loads on the MT specimens are shown in Fig. 35. About one half of the participants predicted failure loads within 10% of experimental loads. The LEFM methods (Participants 1 and 2), Equivalent Energy (Participant 3), and some applications of the effective K_R-curve with a limit-load condition (Participant 15) had some difficulty. The success of some of the other effective K_R-curve methods (Participants 13 and 14) was due to selecting a good estimate of the limit-load condition. The extremely low predictions from Participant 16 were due to using the wrong equation for the MT specimen. The TPFC (Participant 5) and the K_R-curve with the Dugdale model (Participant 10) had the lowest standard errors (see Table 14).

Three-Hole-Crack Tension Specimens—The range and mean of the ratio of predicted-to-experimental failure loads on the THT specimens are shown in Fig. 36. Here many of the participants overpredicted the failure loads. The finite-element analysis using the CTOD criterion (Participant 17), the TPFC (Participant 4), and the DPFAD (Participant 6) had the lowest standard errors.

Some of the methods are compared in Fig. 37 on how well they can predict failure load as a function of initial crack length for the THT specimen. The LEFM method (Participant 1), the Theory of Ductile Fracture (Participant 8), and the effective K_R-curve with a limit-load condition (Participant 13) showed large errors in predicting the failure loads. The participant using the three-dimensional finite-

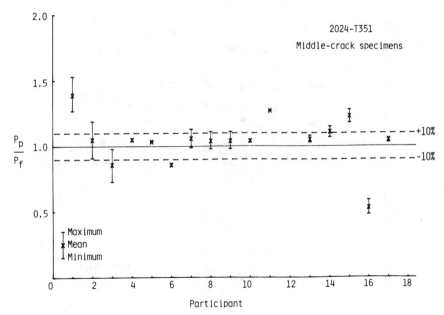

FIG. 35—*Range of ratio of predicted to experimental failure loads on 2024-T351 aluminum alloy middle-crack specimens (four tests).*

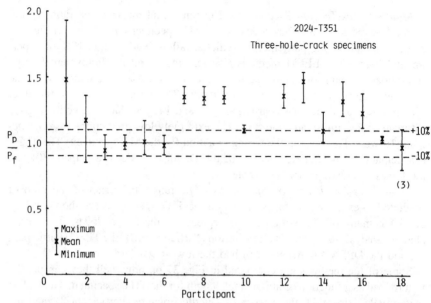

FIG. 36—*Range of ratio of predicted to experimental failure loads on 2024-T351 aluminum alloy three-hole-crack specimens (eight tests; number in parentheses denotes predictions submitted).*

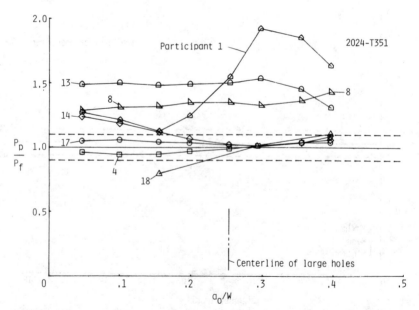

FIG. 37—*Ratio of predicted to experimental failure load as a function of crack length for 2024-T351 three-hole-crack specimens for some selected participants.*

element method with stationary cracks (Participant 18) made only three predictions. A summary of the standard errors is given in Table 14.

Stainless Steel 304

Compact Specimens—The range and mean of the ratio of calculated-to-experimental failure loads on the baseline compact specimens are shown in Fig. 38. Again, most methods were able to correlate the fracture data within ±10%. Participant 13 using a limit-load analysis correlated the results within 20%.

Many of the methods were also able to predict failure loads on the other compact specimens (a_0/W = 0.3 and 0.7) within 10% (Fig. 39). The LEFM method (Participant 1), the DPFAD (Participant 6), the limit-load analysis (Participant 13), and the three-dimensional finite-element method (Participant 18) had some difficulty, but the predictions were generally within 20%. A summary of standard errors is given in Table 15.

Middle-Crack Tension Specimens—A comparison of predictions made on MT specimens are shown in Fig. 40. Here only five participants out of 14 were able to predict failure loads within about 10% of the experimental failure loads. The DPFAD (Participant 6), limit-load analyses (Participants 13, 14, and 15) and the finite-element analysis with the CTOD criterion (Participant 17) were the best methods. With the exception of Participant 16, all other participants overpredicted

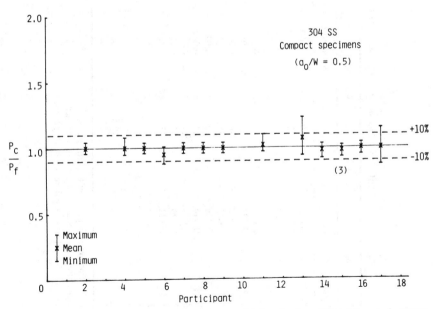

FIG. 38—*Range of ratio of calculated to experimental failure loads on 304 stainless steel compact specimen data supplied to participants (15 tests; number in parentheses denotes calculations submitted).*

TABLE 15—*Ranking of participants in order of minimum standard error on 304 stainless steel.*

Compact[a]		Compact		MT		THT	
Participant	Standard Error	Participant	Standard Error	Participant	Standard Error	Participant	Standard Error
8	0.023	9	0.025	6	0.059	15	0.061
9	0.023	7	0.025	14	0.061	4	0.087
16	0.023	5	0.030	13	0.081	17	0.094[b]
7	0.023	8	0.035	15	0.107	18	0.140
2	0.023	11	0.042	17	0.108	6	0.160
5	0.023	4	0.052	1	0.153	16	0.163
14	0.032	16	0.057	4	0.189	14	0.169
4	0.042	15	0.070	3	0.207	3	0.170
11	0.046[b]	2	0.073	16	0.260	5	0.183
15	0.046	14	0.087	5	0.270	13	0.201
6	0.064	17	0.087	2	0.386	8	0.260
17	0.096	6	0.120	8	0.438	7	0.274
13	0.122	13	0.161	9	0.438	9	0.275
1	...[c]	1	0.167	7	0.440	2	0.378
3	...[c]	18	0.211[b]	11	...[c]	1	0.465
10	...[c]	3	...[c]	10	...[c]	10	...[c]
12	...[c]	10	...[c]	12	...[c]	11	...[c]
18	...	12	...[c]	18	...[c]	12	...[c]

[a]Baseline fracture data provided to participants.
[b]Participant did not predict failure on all specimens.
[c]Participant did not predict failure on these specimens.

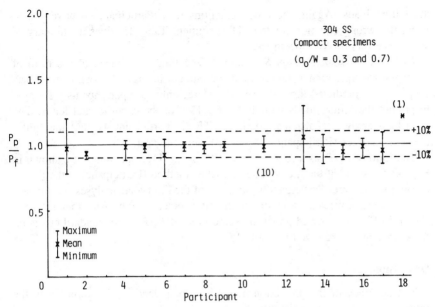

FIG. 39—*Range of ratio of predicted to experimental failure loads on 304 stainless steel compact specimens (twelve tests; number in parentheses denotes predictions submitted).*

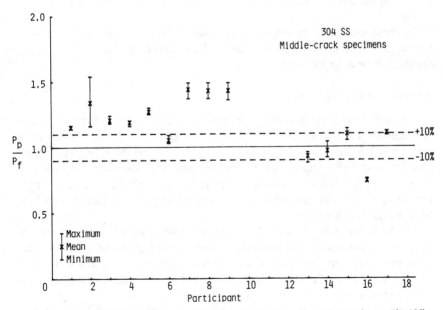

FIG. 40—*Range of ratio of predicted to experimental failure loads on 304 stainless steel middle-crack specimens (four tests).*

the failure loads. Again, the low predictions from Participant 16 were due to using the wrong equation for the MT specimen. Table 15 gives a summary of standard errors on MT specimens.

Three-Hole-Crack Tension Specimens—The range and mean of the ratio of predicted-to-experimental failure loads are shown in Fig. 41. Again, most participants overpredicted the failure loads. Here, only four participants were able to predict the failure loads within about $\pm 15\%$ of the experimental loads. The limit-load analysis (Participant 15), the TPFC (Participant 4), and the finite-element analyses (Participants 17 and 18) were the best.

The ratio of P_p/P_f as a function of a_0/W for the THT specimens are shown in Fig. 42 for some of the methods. The LEFM method (Participant 2), the theory of Doctile Fracture (Participant 9), and one of the limit-load analyses (Participant 16) showed significant variations with initial crack length. All other methods shown in Fig. 40 showed about the same ratio of P_p/P_f. The standard errors on these specimens are given in Table 15.

Discussion

The accuracy of the prediction methods were judged by the variation in the ratio of predicted-to-experimental failure loads; and the predictions were ranked in order of minimum standard error. A summary of the rankings on each material is given in Table 16. These standard errors are the average standard errors on compact ($a_0/W = 0.3$ and 0.7), middle-crack, and three-hole-crack tension specimens. Again, only Participants 4 and 17 knew the failure loads on these specimens.

The best methods were classified as those which had predicted failure loads within $\pm 20\%$ of the experimental failure loads and those that had an average standard error of less than about 0.1.

Aluminum Alloy 7075-T651

For 7075-T651 aluminum alloy, the best methods were the effective K_R-curve (Participants 11–15), the critical crack-tip-opening displacement (CTOD) criterion using a finite-element analysis (Participant 17), and the K_R-curve with the Dugdale model (Participant 10). On the MT and THT specimens, the mean of the ratio of predicted-to-experimental failure loads for all these methods ranged from 0.85 to 0.95. The reason for the slightly lower predicted failure loads from these methods is not clear.

The K_R-curve methods were quite simple to apply for this material because only the stress-intensity factor solutions were required to predict failure of other crack configurations. However, in cases where large amounts of crack extension were required to predict failure, such as in the THT specimens, extrapolation of the K_R-curves was necessary. Participants 12–15 used such extrapolations but Participant 11, who did not make all predictions on the THT specimens, did not use the K_R-curve beyond the data supplied from the baseline compact specimens (see Appendix XI).

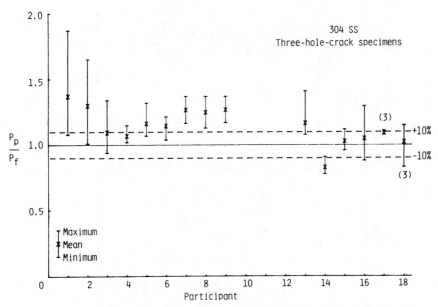

FIG. 41—*Range of ratio of predicted to experimental failure loads on 304 stainless steel three-hole-crack specimens (eight tests; number in parentheses denotes predictions submitted).*

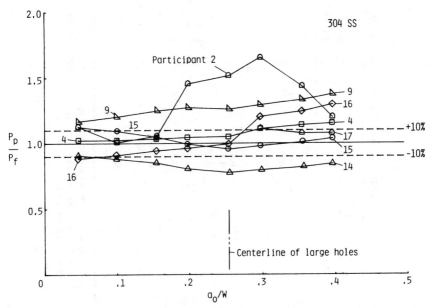

FIG. 42—*Ratio of predicted to experimental failure load as a function of crack length for 304 stainless steel three-hole-crack specimens for some selected participants.*

TABLE 16—Ranking of participants in order of minimum standard error on the three materials for all predictions.

7075-T651		2024-T351		304	
Participant	Standard Error	Participant	Standard Error	Participant	Standard Error
11	0.072[a]	4	0.043	11	0.042[b]
15	0.088	11	0.046[b]	15	0.079
17	0.089	17	0.047	17	0.096[a]
14	0.104	5	0.054	14	0.106
13	0.108	10	0.063	4	0.109
12	0.113[a]	14	0.100	6	0.113
10	0.117	6	0.100	13	0.148
18	0.141[a]	18	0.117[a]	16	0.160
6	0.148	3	0.138[a]	5	0.161
1	0.166	9	0.162	18	0.176[a]
2	0.221	8	0.163	3	0.189[b]
4	0.225	7	0.165	8	0.244
5	0.247	2	0.168	9	0.246
16	0.255	13	0.190	7	0.246
3	0.386[a]	15	0.198	1	0.262
9	0.597	16	0.259	2	0.279
7	0.611	12	0.364[a]	10	...[c]
8	0.637	1	0.380	12	...[c]

[a] Participant did not predict failure on all specimens.
[b] Participant submitted predictions on only the compact specimens.
[c] Participant did not predict failure on this material.

Although the CTOD criterion with the finite-element method was one of the better methods, it required a two-dimensional elastic-plastic finite-element computer program and a large computer system. This method also had some difficulty in predicting the failure loads on the THT specimens for crack-length-to-width (a_0/W) ratios less than 0.2. The largest error was 18%. These large errors may have been due to the large element sizes and approximations used for a_0/W ratios lower than 0.25 (see Ref 20).

Surprisingly, the LEFM methods (Participants 1, 2 and 7–9) and methods which reduce to LEFM for brittle materials (Participants 3–5) had extreme difficulty in predicting failure loads on this material. The 7075-T651 aluminum alloy was selected because it was a low-toughness "brittle" material and LEFM concepts were expected to work well. The calculated plastic-zone sizes (ρ) at failure for this material were small compared to initial crack lengths (ρ/a_0 was less than 0.1). The author suspects that the difficulty arises because of the stable crack growth behavior of this material. Usually, a "brittle" material exhibits very little stable crack growth before failure. However, the stable crack growth behavior of this material was similar to that observed on the 2024-T351 aluminum alloy material (see Figs. 7 and 19). The large compact specimen made of 7075-T651 material had about 7.3 mm of crack extension at maximum load, whereas the 2024-T351 compact specimen had about 6 mm of crack extension at maximum load (see Tables 17 and 18). This may explain why the K_R-curve methods and the finite-element method, which account for stable crack growth, are able to predict the failure loads on this material.

After the initial comparison of the various predictions, Participant 6 rechecked his predictions on the 7075-T651 aluminum MT and THT specimens. He found that he had used incorrect K_R-curves. The corrected calculations gave standard errors of 0.12 for the MT specimens and 0.11 for the THT specimens. These standard errors are consistent with Participant 15, who used methods similar to the one used by Participant 6 (see Table 13).

Aluminum Alloy 2024-T351

For the 2024-T351 aluminum alloy, the best methods (in order of minimum standard error) were the Two-Parameter Fracture Criterion (Participants 4 and 5), the CTOD criterion using the finite-element analysis (Participant 17), the K_R-curve with the Dugdale model (Participant 10), the effective K_R-curve with a limit-load condition (Participant 14), and the Deformation-Plasticity Failure-Assessment Diagram (Participant 6). Participant 11, who used an estimated effective K_R-curve, was not included here because he had made predictions on only some of the compact specimens, and he did not make any predictions on MT and THT specimens.

The TPFC is one of the simplest fracture-analysis methods available for elastic-plastic fracture. Once the two fracture parameters (K_F and m) have been determined for a given material and thickness, closed-form equations are available to

TABLE 17—Typical load, crack-surface displacements, effective crack length, and physical crack length measurements made on 7075-T651 aluminum alloy compact specimens tested at Westinghouse [5].

W, mm	P, kN	V_o, mm	V_{LL}, mm	$(a_e/W)_o$	$(a_e/W)_{LL}$	Compliance $(a_e/W)_o$	$(a_e/W)_{LL}$	Visual $(a/W)_v$
203[a]	16.56	1.04	0.74	0.505	0.504	0.509	0.504	0.503
	21.17	1.45	1.01	0.524	0.516	0.523	0.512	0.516
	22.91	1.65	1.17	0.535	0.529	0.530	0.514	0.528
	24.05	1.91	1.32	0.554	0.543	0.547	0.531	0.537
	23.31	1.99	1.42	0.570	0.562	0.558	0.553	0.554
	23.45	2.19	1.56	0.587	0.579	0.577	0.562	0.567
	22.18	2.32	1.68	0.608	0.601	0.617	0.583	0.589
	20.24	2.48	1.82	0.636	0.629	0.622	0.610	0.617
	16.63	2.74	2.04	0.638	0.676	0.664	0.655	0.664
	14.83	2.88	2.17	0.706	0.700	0.686	0.676	0.689
	12.42	3.07	2.36	0.737	0.733	0.716	0.708	0.727
	9.48	3.25	2.55	0.774	0.771	0.756	0.752	0.764
102[b]	8.38	0.57	0.37	0.513	0.514	0.511	0.504	0.509
	12.88	0.90	0.57	0.520	0.516	0.521	0.521	0.514
	14.18	1.17	0.79	0.555	0.560	0.543	0.541	0.539
	12.73	1.53	1.07	0.627	0.630	0.601	0.604	0.572
	11.03	1.70	1.23	0.668	0.672	0.645	0.644	0.589
	8.78	1.85	1.37	0.713	0.718	0.689	0.692	0.659
	6.19	2.32	1.57	0.768	0.772	0.748	0.748	0.716
	4.04	2.41	1.87	0.822	0.820	0.803	0.804	0.766
51[c]	8.23	0.53	0.35	0.513	0.505	0.508	0.509	0.513
	8.23	0.64	0.44	0.554	0.547	0.539	0.548	0.523
	8.18	0.78	0.56	0.595	0.594	0.571	0.572	0.568
	8.15	0.90	0.65	0.622	0.618	0.594	0.586	0.608
	6.98	1.06	0.76	0.673	0.668	0.636	0.636	0.648
	5.33	1.15	0.88	0.723	0.723	0.685	0.689	0.683
	4.14	1.30	0.99	0.767	0.764	0.728	0.733	0.713
	2.99	1.38	1.07	0.805	0.804	0.766	0.769	0.783
	2.37	1.51	1.17	0.832	0.830	0.795	0.797	0.808

[a] $B = 12.68$ mm and $a_0/W = 0.503$.
[b] $B = 12.80$ mm and $a_0/W = 0.509$.
[c] $B = 12.75$ mm and $a_0/W = 0.508$.

predict failure loads on other configurations. Participant 5, who used a modified version of the TPFC (see Appendix VII), did not know the failure loads on any of the specimens except the baseline compact specimens. Although Participant 5 had very good predictions on the THT specimens, he did not, however, analyze this specimen as recommended in the TPFC analysis. He treated this specimen like a MT specimen with the "crack length plus hole diameter" as the total crack length in a MT specimen.

Again, the finite-element analysis with the CTOD criterion predicted the stable crack growth behavior (Participant 17) and the failure loads extremely well on this material.

The K_R-curve with the Dugdale model (Participant 10) was also very simple to apply. However, the K_R-curves for 2024-T351 compact specimens were different for different specimen sizes (see Fig. 19). This variation in the K_R-curve with specimen size could lead to errors in predicting failure loads.

The effective K_R-curve with a limit-load condition made good predictions on this material because a good value of limit load was selected. As pointed out by Participant 15 (Appendix XV, Fig. 56), the effective K_R-curve concept cannot be applied to this material because the K_R-curve for the tension specimen appears to be significantly lower than that obtained from the compact specimens.

The Deformation-Plasticity Failure-Assessment Diagram (Participant 6) also made good predictions on this material. This method was the only one to use the J_R-curve supplied on the baseline compact specimens. The J_R-curve was used to develop the deformation-plasticity failure-assessment diagram for each material (see Appendix VIII).

Stainless Steel 304

For 304 stainless steel, the best methods were the limit-load criteria (Participants 14 and 15), the CTOD criterion using the finite-element analysis (Participant 17), the TPFC (Participant 4), and the Deformation Plasticity Failure-Assessment Diagram (Participant 6). Again, Participant 11 was not included here because he did not make any predictions on MT and THT specimens. With the exception of Participant 14, all of these methods tended to overpredict the failure loads on MT and the THT specimens. The mean of the ratio of predicted-to-experimental failure loads ranged from 1.03 to 1.2. However, Participant 14 had mean values ranging from 0.8 to 0.93.

Participants 14 and 15 used the same method—a limit-load analysis—but each participant selected a different value of flow stress (σ_o). For the large MT and THT specimens, Participant 15 chose σ_o as 480 MPa and Participant 14 chose σ_o as 390 MPa. Participant 15 tended to slightly overpredict failure loads and Participant 14 tended to slightly underpredict failure loads. The specimens, however, tended to fail at a flow stress of 450 MPa based on the initial net-section area. This stress is nearly the average between the yield stress and the ultimate

TABLE 18—*Typical load, crack-surface displacements, effective crack length, and physical crack length measurements made on 2024-T351 aluminum alloy compact specimens tested at Westinghouse [5].*

W, mm	P, kN	V_o, mm	V_{LL}, mm	Compliance					Visual $(a/W)_x$
				$(a_e/W)_o$	$(a_e/W)_{LL}$	$(a/W)_o$	$(a/W)_{LL}$		
203^a	13.18	0.84	0.57	0.504	0.503	0.508	0.507		0.504
	21.64	1.41	0.95	0.509	0.507	0.510	0.500		0.505
	28.28	1.87	1.28	0.512	0.513	0.506	0.501		0.506
	36.30	2.51	1.71	0.522	0.522	0.511	0.510		0.507
	42.13	3.05	2.12	0.533	0.534	0.513	0.510		0.512
	48.32	3.82	2.65	0.551	0.551	0.515	0.513		0.516
	51.48	4.51	3.18	0.572	0.572	0.524	0.522		0.525
	52.02	5.18	3.71	0.596	0.597	0.530	0.536		0.554
	51.35	5.70	4.11	0.616	0.617	0.537	0.543		0.573
	48.05	6.38	4.67	0.647	0.648	0.557	0.561		0.598
	42.44	7.03	5.19	0.681	0.681	0.583	0.582		0.629
	38.57	7.58	5.65	0.706	0.707	0.604	0.606		0.648
	32.82	8.24	6.20	0.737	0.737	0.607	0.638		0.679
	28.55	8.78	6.70	0.761	0.761	0.658	0.662		0.704

102[b]	8.23	0.56	0.39	0.510	0.516	0.516	0.513	0.512
	15.92	1.14	0.76	0.520	0.518	0.508	0.499	0.512
	23.22	1.85	1.23	0.543	0.541	0.504	0.499	0.520
	27.33	2.53	1.75	0.573	0.574	0.513	0.508	0.532
	28.75	3.18	2.24	0.607	0.609	0.522	0.523	0.552
	25.91	3.99	2.90	0.667	0.666	0.544	0.554	0.562
	21.80	4.64	3.45	0.710	0.713	0.576	0.583	0.582
	18.37	5.11	3.87	0.744	0.747	0.608	0.618	0.630
	14.69	5.72	4.36	0.781	0.783	0.647	0.656	0.667
51[c]	6.21	0.42	0.28	0.513	0.516	0.525	0.526	0.522
	11.44	0.83	0.59	0.536	0.542	0.517	0.521	0.522
	13.98	1.24	0.87	0.577	0.576	0.525	0.528	0.527
	14.77	1.63	1.17	0.618	0.618	0.535	0.539	0.537
	13.80	2.21	1.52	0.679	0.669	0.556	0.562	0.555
	12.58	2.47	1.83	0.708	0.707	0.569	0.577	0.617
	10.62	2.92	2.20	0.750	0.749	0.607	0.613	0.632
	8.25	3.34	2.53	0.791	0.790	0.648	0.648	0.657
	6.21	3.86	2.97	0.828	0.828	0.690	0.693	0.717

[a]B = 12.60 mm and a_0/W = 0.504.
[b]B = 12.52 mm and a_0/W = 0.512.
[c]B = 12.55 mm and a_0/W = 0.522.

tensile strength. Thus, a proper selection of the flow stress makes this method highly desirable because of its simplicity ($P = \sigma_o A_{net}$).

Surprisingly, the finite-element analysis with the CTOD criterion predicted failure loads on the steel specimens quite well (within $\pm 10\%$ of the experimental failure loads). The analysis was based on a "small strain" assumption and the steel specimens failed under massive plastic deformation (see Fig. 14c). An observation of Fig. 14c would imply that a finite-deformation analysis should have been used on this material. It is believed that the small strain analysis is being tuned by fitting to the baseline compact specimen data. And this, in turn, is enabling the analysis to predict failure loads on the other specimens. Participant 17, however, did not make all predictions on the THT specimens because of the high computer cost. He did not make predictions for a_0/W ratio less than or equal to 0.25. It would have been interesting to see if this method could have predicted failure loads on the specimens with the small initial crack lengths.

Again, the TPFC analysis (Participant 4) worked reasonably well on the ductile material. This method tended to overpredict failure loads. On the MT specimens, predicted failure loads were 15 to 20% higher than experimental loads. But on the THT specimens, the predicted failure loads ranged from 2 to 15% higher than the experimental loads.

The Deformation Plasticity Failure-Assessment Diagram (Participant 6) was able to predict failure loads on compact, MT, and THT specimens generally within $\pm 25\%$ of experimental failure loads. The application of this method to a highly strain-hardening material demonstrates why strain-hardening properties must be taken into account. The original failure assessment diagram (or R-6 approach) would not have been able to predict failure loads very accurately on this material.

Concluding Remarks

An experimental and predictive round robin was conducted by the American Society for Testing and Materials Task Group E24.06.02 on Application of Fracture Analysis Methods. The objective of the round robin was to verify whether fracture analysis methods currently used could predict failure loads on cracked structural components from results on compact specimens. Results of fracture tests conducted on various-size compact specimens made of 7075-T651 aluminum alloy, 2024-T351 aluminum alloy, and 304 stainless steel were supplied as baseline data to 18 participants. Tensile stress-strain properties on full-thickness specimens were also provided. These participants used 13 different fracture analysis methods to predict failure loads on other compact specimens, middle-crack tension specimens, and structurally configured specimens. This specimen, containing three circular holes with a crack emanating from one of the holes, was subjected to tensile loading.

The accuracy of the prediction methods was judged by the variations in the ratio of predicted-to-experimental failure loads. The range of applicability of the prediction methods was also considered in assessing their usefulness. The best methods were judged to be those which had predicted failure loads within $\pm 20\%$ of experimental failure loads and those that had an average standard error of less than about 0.1 in the ratio of predicted-to-experimental failure loads. For each material, these methods were ranked in order of minimum standard error.

ALUMINUM ALLOY 7075-T651

1. Effective K_R-Curve (Participant 11)
2. Effective K_R-Curve (Participant 15)
3. Finite-Element Analysis with Critical CTOD Criterion (Participant 17)
4. Effective K_R-Curve (Participant 14)
5. Effective K_R-Curve (Participant 13)
6. K_R-Curve with Dugdale Model (Participant 10)

ALUMINUM ALLOY 2024-T351

1. Two-Parameter Fracture Criterion (Participant 4)
2. Finite-Element Analysis with Critical CTOD Criterion (Participant 17)
3. Modified Two-Parameter Fracture Criterion (Participant 5)
4. K_R-Curve with Dugdale Model (Participant 10)
5. Effective K_R-Curve with Limit-Load Criterion (Participant 14)
6. Deformation Plasticity Failure Assessment Diagram (Participant 6)

STAINLESS STEEL 304

1. Limit-Load Criterion (Participant 15)
2. Finite-Element Analysis with Critical CTOD Criterion (Participant 17)
3. Limit-Load Criterion (Participant 14)
4. Two-Parameter Fracture Criterion (Participant 4)
5. Deformation Plasticity Failure Assessment Diagram (Participant 6)

From the results of the experimental and predictive round robin, many of the fracture-analysis methods tried could predict failure loads on several crack configurations for a wide range in material behavior. In several cases, the analyst had to select the method he thought would work the best. This would require experience and engineering judgment. Some methods, however, could be applied to all crack configurations and materials considered. Many of the large errors in predicting failure loads were due to improper application of the method or human error. As a result of the round robin, many improvements have been made in these and other fracture-analysis methods.

Acknowledgment

The author would like to thank M. R. Gardner and R. Cherry of the NASA Langley Research Center for their superb efforts in testing the large number of fracture specimens required for the ASTM E24.06.02 Round Robin on Fracture.

APPENDIX I

by J. C. Newman, Jr.

Stress-Intensity Factors for the Three-Hole-Crack Tension Specimen

The stress-intensity factor solution for the three-hole-crack tension specimen (Fig. 1c) was obtained by using a two-dimensional elastic finite-element analysis under plane-stress conditions. The stress-intensity factors were obtained by using a local-energy approach [22]. To verify the finite-element approximation of the local-energy approach and the mesh pattern in the crack-tip region, compact and middle-crack tension specimens were also analyzed (see Ref 20).

The finite-element mesh used for the three-hole-crack tension (THT) specimen is shown in Fig. 43. The smallest element size along the crack line was 0.00625 times the specimen width (W). The number of elements (constant strain) and number of nodes are shown in the figure.

The stress-intensity factors for the THT specimen were calculated for various crack-length-to-width (a/W) ratios and the results are shown in Fig. 44 as symbols. Stress-intensity factors were normalized by gross applied stress (P/WB) and are plotted against a/W with $W = 254$ mm. Crack length is measured from the edge of the small hole. The dash-dot line shows the center-line of the two large holes. On the basis of the results obtained on the compact and middle-crack tension specimens [20], the finite-element results are expected to be about 1% to 3% lower than the "exact" stress-intensity factor solution.

For convenience, an equation was fit to the finite-element results and is

$$K = \frac{P}{WB} \sqrt{\pi a}\, F \tag{20}$$

where

$$F = \sum_{i=1}^{4} \sum_{j=1}^{2} \frac{A_{ij}\, (1 - a/b)^{-1/2}}{(1 + a/r)^{i-1} \sqrt{[(y_o/x_o)^2 + (a/x_o - 1)^2]^{j-1}}} \tag{21}$$

$$A_{11} = 2.02 \qquad A_{12} = -9.17$$
$$A_{21} = -62.37 \qquad A_{22} = 287.72$$
$$A_{31} = 1025.8 \qquad A_{32} = -2845.1$$
$$A_{41} = -8270.6 \qquad A_{42} = 11\,927.3$$

and $r = 12.7$ mm (0.5 in.), $b = 165$ mm (6.5 in.), $x_o = 63.5$ mm (2.5 in.), and $y_o = 50.8$ mm (2 in.). Equation 20 is believed to be within ±2% of the exact solution and is shown as the curve in Fig. 44.

APPENDIX II

by D. E. McCabe

Baseline Compact Specimen Fracture Data

This Appendix presents data generated on compact specimens made of the three materials and tested by Westinghouse Research Laboratory [5]. The compact specimen configuration was tested with three specimen widths ($W = 51$, 102, and 203 mm) but the nominal thickness of all specimens was 12.7 mm. The nominal crack-length-to-width ratio (a_0/W)

Mesh A: Elements = 3842; Nodes = 2077

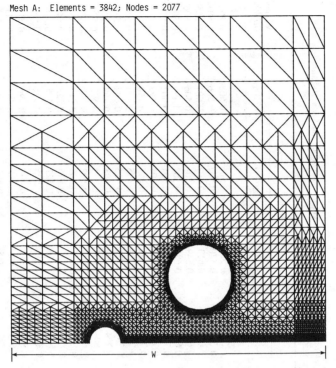

FIG. 43—*Finite-element idealization of one half of the three-hole-crack specimen.*

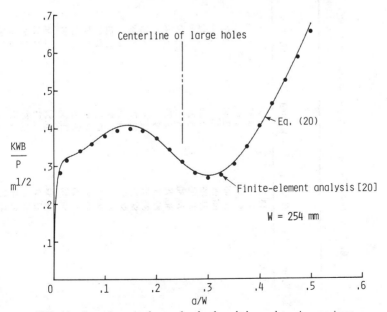

FIG. 44—*Stress-intensity factors for the three-hole-crack tension specimen.*

TABLE 19—Typical load, crack-surface displacements, effective crack length, and physical crack length measurements made on 304 stainless steel compact specimens tested at Westinghouse [5].

W, mm	P, kN	V_o, mm	V_{LL}, mm	Compliance				Visual $(a/W)_y$
				$(a_e/W)_o$	$(a_e/W)_{LL}$	$(a/W)_o$	$(a/W)_{LL}$	
203[a]	13.36	0.30	0.20	0.505	0.501	...	...	...
	33.67	0.85	0.56	0.532	0.524	0.506	0.502	0.502
	50.33	1.56	1.07	0.574	0.570	0.507	0.506	0.503
	60.48	2.40	1.65	0.620	0.613	0.507	0.501	0.505
	66.09	3.23	2.26	0.655	0.649	0.505	0.506	0.506
	71.08	4.39	3.12	0.691	0.686	0.505	0.499	0.509
	73.93	5.28	3.79	0.711	0.707	0.504	0.502	0.510
	78.30	6.95	5.02	0.739	0.735	0.501	0.499	0.513
	80.79	8.29	6.04	0.756	0.752	0.495	0.498	0.516
	83.20	9.91	7.24	0.772	0.768	0.495	0.494	0.520
	85.07	11.30	8.27	0.784	0.779	0.495	0.494	0.523
	86.31	12.78	9.38	0.794	0.790	0.498	0.498	0.526
	86.85	14.33	10.55	0.804	0.800	0.500	0.502	0.530
	86.85	15.60	11.51	0.811	0.808	0.502	0.507	0.532
	86.49	17.35	12.87	0.821	0.818	0.509	0.516	0.537
	85.87	18.86	13.97	0.828	0.825	0.515	0.523	0.541
	84.98	20.22	15.04	0.834	0.832	0.521	0.525	0.546
	84.18	21.34	15.91	0.839	0.837	0.525	0.531	0.561
102[b]	19.84	0.57	0.36	0.560	0.547	0.498	0.494	0.510
	33.31	1.71	1.13	0.664	0.655	0.498	0.494	0.511
	37.48	2.88	1.96	0.721	0.715	0.502	0.494	0.515
	41.40	4.53	3.15	0.763	0.759	0.506	0.494	0.519
	43.11	5.61	3.95	0.782	0.778	0.490	0.494	0.522
	45.07	7.12	5.04	0.800	0.797	0.494	0.494	0.526

Specimen								
	46.29	8.37	5.98	0.812	0.810	0.490	0.504	0.529
	47.18	9.31	6.68	0.820	0.817	0.486	0.499	0.531
	47.76	10.64	7.68	0.830	0.828	0.498	0.499	0.535
	47.62	11.86	8.55	0.838	0.836	0.510	0.529	0.539
	47.47	13.09	9.42	0.845	0.843	0.510	0.519	0.545
	46.78	14.77	10.69	0.854	0.853	0.521	0.524	0.553
	46.69	15.57	11.30	0.858	0.857	0.521	0.541	0.558
	44.87	17.28	12.57	0.868	0.886	0.524	0.550	0.568
	43.85	18.31	13.39	0.872	0.871	0.524	0.550	0.574
51[c]	9.80	0.25	0.18	0.541	0.543	0.522	0.505	—[d]
	16.58	0.77	0.55	0.653	0.645	0.515	0.519	—[d]
	18.87	1.37	0.98	0.719	0.709	0.511	0.528	—[d]
	20.36	2.01	1.48	0.756	0.750	0.515	0.528	—[d]
	21.72	2.64	1.94	0.778	0.772	0.507	0.519	—[d]
	22.72	3.15	2.35	0.791	0.786	0.491	0.505	—[d]
	23.66	3.91	2.93	0.808	0.803	0.491	0.509	—[d]
	24.45	4.63	3.49	0.819	0.815	0.495	0.514	—[d]
	25.45	5.57	4.24	0.831	0.827	0.499	0.519	—[d]
	25.94	6.29	4.80	0.839	0.835	0.511	0.509	—[d]
	26.27	6.99	5.35	0.846	0.842	0.507	0.519	—[d]
	26.69	7.69	5.92	0.851	0.848	0.507	0.519	—[d]
	26.76	8.31	6.42	0.856	0.854	0.511	0.519	—[d]
	26.47	9.81	7.27	0.865	0.862	0.511	0.528	—[d]
	26.14	10.12	7.91	0.870	0.869	0.522	0.536	—[d]
	25.65	10.69	8.38	0.874	0.873	0.522	0.540	—[d]

[a]B = 12.78 mm and a_0/W = 0.502.
[b]B = 13.03 mm and a_0/W = 0.510.
[c]B = 12.80 mm and a_0/W = 0.518.
[d]Visual crack lengths were not measured.

was 0.5. Crack-opening displacements were measured at two locations (crack mouth, V_o, and load line, V_{LL}) along the crack plane. A two-pen type recorder was used to plot V_o and V_{LL} displacement against applied load. All specimens were fatigue precracked according to ASTM E 399 requirements. During the fracture test (under displacement control), all specimens were periodically partially unloaded (about 15%) to obtain unloading slopes (dP/dV) which were used to determine the amount of stable crack growth, Δa physical (Δa_p). In addition, the crack was observed visually to the nearest 1 mm on the specimen surfaces. Such optical measurements are highly imprecise, especially with plastic-zone interference and crack tunneling. Representative load-displacement records and crack extension measurements for each material and specimen size are contained herein. Crack-growth resistance (K_R) curves were developed in terms of Δa physical and Δa effective crack extension for each material and specimen size.

Typical Data Sets

Twenty-seven specimens were tested (nine specimens per material) with a nominal a_0/W of 0.5. Tables 17–19 present typical data for each material and specimen width. These tables give applied load, displacement at V_o and V_{LL} locations, effective (a_e) and physical (a) crack length determined from V_o and V_{LL} compliance data, and visual crack lengths measured on the specimen surfaces. The data listed are for each unloading point on the test record.

Crack-Extension Measurements

Physical crack extensions (Δa_p) used in determining K_R and J_R data were obtained from unloading compliance for the two aluminum alloys. This was justified by a $\pm 5\%$ correlation with crack lengths measured on the specimen surfaces (visual). Crack extensions determined from compliance on the stainless steel specimens had equal precision but the comparison with visual crack lengths was unacceptable because of the large deformations involved in the tests. Therefore, the Δa_p values used in K_R and J_R calculations were visual measurements.

Failure Load Data

Tables 5–7 include the failure loads obtained on the baseline compact specimens tested by Westinghouse.

APPENDIX III

by D. P. Peng

Critical Stress-Intensity Factor Approach (Participant 1)

The fracture-analysis method used was a critical elastic stress-intensity factor approach accounting for width effects. The method consists of three steps. First, the critical values of stress-intensity factor, K_{Ie}, are calculated using the initial crack lengths and failure loads on the baseline compact specimen data provided. For each material, there were three groups of K_{Ie}, one group for $W = 51$ mm, one group for $W = 102$ mm, and one group for $W = 203$ mm. The average K_{Ie} for each group was calculated and these values are shown in Fig. 45 as a function of W. A "K_{Ie} against W" curve was then generated by curve-fitting on the average values (solid curves in Fig. 45). Second, the stress-intensity boundary-correction factors for the compact, middle-crack, and three-hole-crack specimens were determined. Third, the predicted failure loads were then calculated by using a K_{Ie} value interpolated or extrapolated from the appropriate material curve at the specified

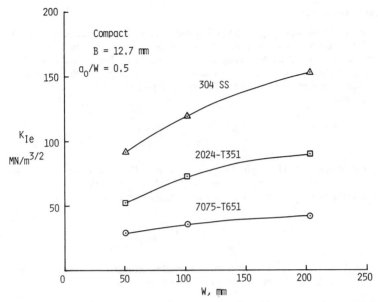

FIG. 45—*Critical elastic stress-intensity factor as a function of compact specimen width for the three materials.*

specimen width (W) and the stress-intensity factors determined in the second step. Failure was assumed to occur when K was equal to K_{Ie}. The failure loads were given by

$$\text{Compact:} \quad P = K_{Ie}\, B\sqrt{W}/F(a_0/W) \tag{22}$$

$$\text{MT:} \quad P = K_{Ie}\, WB/\sqrt{\pi a_0\, \sec\,(\pi a_0/W)} \tag{23}$$

$$\text{THT:} \quad P = K_{Ie}\, WB/[\sqrt{\pi a_0}\, F(a_0/W)] \tag{24}$$

where F in equations (22) and (24) is the appropriate boundary-correction factor.

APPENDIX IV

by T. M. Hsu

Plastic-Zone Corrected Stress-Intensity Factor Approach (Participant 2)

The fracture-analysis method used was a critical elastic stress-intensity factor approach accounting for plastic yielding at the crack tip. The analysis was based on the following assumptions:

1. The plastic zone at the crack tip is small compared with crack size (r_p/a is less than 0.1).
2. Fracture will occur when the plasticity-corrected stress-intensity factor reaches the critical value K_c.

Fracture Toughness

The fracture toughness, K_c, is defined as the K value at the onset of fracture. K_c is a material constant for a specific processing sequence, grain orientation, and thickness. It is independent of crack size but varies with specimen width. For a constant thickness, K_c is normally higher for larger values of width. The K_c values used in the predictions were the average values obtained from the baseline compact specimens. The K_c values were computed from

$$K_c = \frac{P_f}{B\sqrt{W}} F\left(\frac{a_e}{W}\right) \tag{25}$$

where the effective crack length, a_e, was $a_0 + r_p$. $F(a_e/W)$ is the usual boundary-correction factor for the compact specimen (ASTM E 399) written in terms of a_e. The plastic-zone size, r_p, was calculated from Irwin's plane-strain equation as

$$r_p = 0.056 \left(\frac{K}{\sigma_{ys}}\right)^2 \tag{26}$$

The average K_c values computed from the baseline compact specimen data for each specimen width are shown in Fig. 46 for the three materials.

Failure Load Predictions

The K_c values shown in Fig. 46 at the appropriate width were used to predict failure loads on the other compact specimens ($a_0/W = 0.3$ and 0.7). For MT and THT specimens, there are no known transfer functions to account for the size effect using compact specimen data. Therefore, an average value of K_c, denoted as $\bar{K}_c$ (dashed line in figure), was used to predict failure loads on these specimens. The following equations were used to predict failure loads

$$\text{Compact:} \quad P = K_c B\sqrt{W}/F(a_e/W) \tag{27}$$

$$\text{MT:} \quad P = \bar{K}_c WB/\sqrt{\pi a_e} \sec (\pi a_e/W) \tag{28}$$

$$\text{THT:} \quad P = \bar{K}_c WB/[\sqrt{\pi a_e} F(a_e/W)] \tag{29}$$

The effective crack length (a_e) was ($a_0 + r_p$) where r_p was calculated from Eq 26 using either K_c or $\bar{K}_c$.

For the THT specimen, $F(a_e/W)$ was given by Eq 21 with crack length (a) replaced by a_e. If the initial crack length was in the range $35 \leq a_0 \leq 76$ mm, however, the running crack would be arrested (ignoring dynamic effects). Therefore, the effective crack length was set equal to 76 mm in the calculation of failure load.

APPENDIX V

by F. J. Witt

Equivalent Energy Method (Participant 3)

The fracture-analysis method used was the Equivalent Energy method. The Equivalent Energy method is essentially a modeling procedure for determining the elastic-plastic

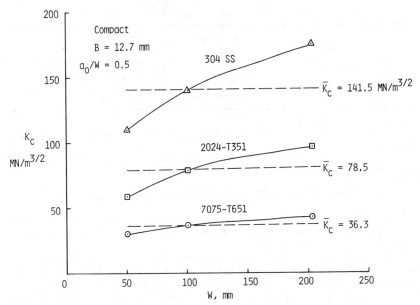

FIG. 46—*Critical fracture toughness (plastic-zone corrected stress-intensity factor) as a function of compact specimen width for the three materials.*

fracture conditions of a flawed structure. Its basis lies in the uniqueness of the volumetric energy ratio as discussed in Ref *10*. From this uniqueness, the governing equation is

$$K_{\mathrm{Icd}} = S_d \sqrt{\pi a_d} \, F_d \tag{30}$$

where

K_{Icd} = critical fracture toughness parameter derived from a "standard" fracture toughness specimen with the thickness equal to that of the structure of interest and based on energy to maximum load,

F_d = shape factor calibrated to the particular stress (S_d), flaw type, and configuration,

S_d = equivalent stress, which is based on the elastic load line or the linear extension thereof, and

a_d = crack length.

The area under the triangle (energy) defined by S_d is the equivalent energy to the maximum load failure condition for the actual load (or stress) displacement (or strain) curve which may define either elastic or elastic-plastic behavior, such behavior being at the flaw location without the flaw being present. Although the most successful applications of the Equivalent Energy method have been in those cases where the shape factors have been determined experimentally, the uniqueness of the volumetric energy ratio implies that F_d is the same as the boundary-correction factor from linear-elastic fracture mechanics.

For the present application, linear elastic shape factors were provided based on plane-strain analyses. Load-displacement data were provided for non-standard compact specimens each with a thickness of approximately 12.7 mm but a plan view corresponding to "standard" 25, 50, and 100-mm-thick compact specimens. Maximum load K values were calculated for each data set provided and linear extrapolation was used to estimate a value of K_{Icd} for a 12.7-mm-thick "standard" compact specimen ($W = 25.4$ mm). These values were used in the analysis and are given in Table 20.

TABLE 20—*Fracture toughness values used in Equivalent Energy method.*

Material	Estimated K_{Icd},[a] MN/m$^{3/2}$	Experimental K_{Icd},[b] MN/m$^{3/2}$
7075-T651[c]	29.7	28.3,[d] 27.9[d] to 31.1[e]
2024-T351	59.6	55.0, 53.1, 57.5
304	423.8	385.1, 392.8

[a]Standard K_{Ic} specimen (W = 25.4 mm).
[b]Specimens (W = 25.4 mm) tested at Westinghouse on the same stock of material as used in the round robin.
[c]The instability-arrest (pop-in) behavior prior to maximum load was ignored in evaluation of maximum-load K values.
[d]Valid K_{Ic} values.
[e]Two values reflect different maximum load selection.

In lieu of having load-displacement curves for the actual specimens, the stress-strain curves obtained from the tensile specimens with square cross sections were used as provided for the MT and THT specimens.

Within the above assumption, the equivalent stresses, hence the equivalent energies, were calculated for all the specimens tested except the compact specimens. Only the equivalent energies were calculated for the compact specimens. Both failure load and failure strains (if the proportional limits were exceeded) were estimated for the MT and THT specimens. Because of the nonmonotonic stress-intensity factor for the THT specimen, the maximum load was taken as the load for a crack length of 81 mm when the initial crack length was less than or equal to 81 mm.

Two further investigations were subsequently carried out. A literature survey yielded test results applicable to the two MT specimens (W = 125 and 250 mm). The strain-hardening behavior for 13-mm-thick tension specimens is somewhat dependent on the specimen width-to-specimen height ratio. An investigation of elastic stress-strain results suggests that the elastic modulus varies considerably across the midsection. Also, two standard 12.7-mm-thick compact specimens were tested of each material. These results are also given in Table 1. The ½T results for the 7075-T651 aluminum alloy were valid K_{Ic} values.

A comparison of predicted loads with experimental loads for the MT and THT specimens revealed the following:

1. For the 7075-T651 aluminum alloy, predicted loads were very conservative, low by a factor of 2 in some cases.
2. For the 2024-T351 aluminum alloy, the predicted loads were generally conservative (about 10% lower than the experimental loads).
3. For the 304 stainless steel, the analysis overpredicted (about 15%) the failure load for nine of the twelve cases.

Because the gross-section stresses on the 7075-T651 aluminum alloy specimens were well below the proportional limit, significant unknown elastic-plastic anomalies from using the stress-strain curves would not be anticipated. If for a constant thickness the plane-strain shape factor is corrected for constraint by the ratio of the K_{Ic} to the maximum load K corresponding to the same length of remaining ligament as the panel specimen, the experimental loads are found within an average of 4% for the middle-crack specimens and within 7% for the three-hole specimens. It would seem that the constraint factors could be estimated by calculational methods such as those of Ref 23. The modulus of elasticity of a material would probably become involved in estimating constraint on an energy basis.

APPENDIX VI

by J. C. Newman, Jr.

Two-Parameter Fracture Criterion (Participant 4)

The fracture-analysis method used was a one-parameter version of the Two-Parameter Fracture Criterion (TPFC). The TPFC is described in Refs *11* and *12* for compact and MT specimens. The TPFC consists of determining two fracture parameters (K_F and m) from fracture data (initial crack lengths and failure loads) on a given specimen type. After the two parameters have been determined, nominal (net-section) failure stresses for other crack configurations were given by

$$S_n = \frac{K_F}{\sqrt{\pi a_0}\, F_n + (mK_F/S_u)} \quad \text{for} \quad S_n < \sigma_{ys} \tag{31}$$

and

$$S_n = \sqrt{(m\gamma)^2 + 2\gamma S_u} - m\gamma \quad \text{for} \quad S_u > S_n \geq \sigma_{ys} \tag{32}$$

where

$$\gamma = \frac{K_F\, \sigma_{ys}}{2S_u\sqrt{\pi a_0}\, F_n} \tag{33}$$

The function F_n is the usual stress-intensity boundary-correction factor with a nominal-stress-to-gross-stress conversion factor (see Ref *11*). S_u is the nominal stress required to produce a fully plastic region (or hinge) on the net section. For the MT specimen S_u is equal to σ_u. For the compact specimen, S_u is a function of load eccentricity and is 1.62 σ_u for an a_0/W ratio of 0.5 [*12*].

If a relationship between the two fracture parameters exists, then the analysis could be simplified; and the elastic-plastic fracture toughness K_F could be expressed directly in terms of failure load, initial crack length, specimen type, and tensile properties (σ_{ys} and σ_u). Such a relationship was observed in Ref *11*. The ratio of fracture toughness to modulus, K_F/E, is plotted as a function of m in Fig. 47 for surface cracks and through cracks in a wide range of materials. The through cracks were in middle-crack, compact, or notch-bend specimens. Each data point indicates K_F/E and m obtained from each set of fracture data analyzed (material and specimen type). The results suggest a common functional relationship between the two fracture parameters for steels, titanium alloys and aluminum alloys and, possibly, magnesium alloys.

The solid curve in Fig. 47 shows the results of an equation

$$m = \tanh\left(21\, \frac{K_F}{E}\right) \tag{34}$$

fit to the experimental data (K_F/E given in mm$^{1/2}$). Equation 34 was used to eliminate m from the fracture criterion. Thus, only one test, in principle, was needed to evaluate K_F. The value of K_F was determined from the baseline compact specimen data using a least-squares regression analysis [*11*] and the corresponding m value was calculated from Eq 34. The fracture parameters are given in Table 9.

The failure stresses (or loads) on compact and on MT specimens were computed from Eqs 31 and 32. The functions F_n are given in Ref *12*. The procedure used to compute failure stresses on THT specimens is described herein.

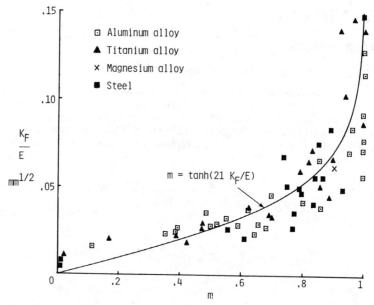

FIG. 47—*Experimental relationship between the two fracture constants* K_F *and* m.

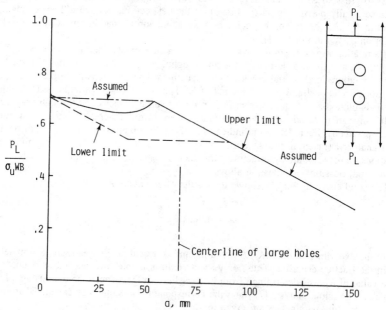

FIG. 48—*Upper and lower limit-load calculations using ultimate tensile strength for three-hole-crack tension specimen.*

To calculate the stress S_u, a limit-load analysis was conducted on the THT specimen. Figure 48 shows the upper and lower bounds using the ultimate tensile strength in a limit-load analysis described in Ref 24. The limit load P_L normalized by $\sigma_u WB$ is plotted against crack length, a. For simplicity, the upper limit was assumed to be $0.7\sigma_u WB$ (dash-dot line) for crack lengths less than 50 mm. The upper limit was used to calculate S_u and was

$$S_u = \sigma_u \quad \text{for} \quad a_0 \geqslant 50 \text{ mm} \tag{35}$$

and

$$S_u = \frac{0.7}{1 - [(D + a_0)/W]} \sigma_u \quad \text{for} \quad a_0 < 50 \text{ mm} \tag{36}$$

Again, Eqs 31 and 32 were used to calculate failure stresses for the THT specimen with S_u given by Eqs 35 and 36; and F_n is given by

$$F_n = \left(1 - \frac{D + a_0}{W}\right) F \tag{37}$$

where F is given by Eq 21. The term preceding F in Eq 37 is the nominal-stress-to-gross-stress conversion factor.

The failure load was then calculated by

$$P = S_n WB \left(1 - \frac{D + a_0}{W}\right) \tag{38}$$

APPENDIX VII

by C. M. Hudson and P. E. Lewis

Modified Two-Parameter Fracture Criterion (Participant 5)

The fracture-analysis method used was a modified version of the Two-Parameter Fracture Criterion [11,12]. The fracture data on the baseline compact specimens given in Tables 5a, 6a, and 7a for a nominal a_0/W ratio of 0.5 were used to obtain the two parameters, K_F and m, for each of the three materials. The fracture parameters are given in Table 10. These values of K_F and m were used in subsequent analyses to predict failure loads on compact, MT, and THT specimens. (Note that the parameter m was not truncated to 1.0, as is recommended [11]).

For all three types of specimens, the elastic nominal (net-section) stress at failure, S_n, was determined from Eqs 31 and 32 in Appendix VI. Reference 12 presents equations for S_u for the compact and MT specimens (see Appendix VI). For the THT specimen, S_u was set equal to σ_u, that is, the same as for the MT specimen. After S_n was calculated, the failure load, P, was determined for all three materials as follows. For the compact specimens

$$P = S_n WB(1 - \lambda) \bigg/ \left(1 + 3\frac{1 + \lambda}{1 - \lambda}\right) \tag{39}$$

where $\lambda = a_0/W$. For the MT specimens

$$P = S_n WB(1 - 2a_0/W) \tag{40}$$

Equations 39 and 40 were obtained from Ref *12*. The THT specimen was treated exactly the same as the MT specimen, except that the initial crack length, $2a_0$, was replaced with the hole diameter plus the initial crack length $(D + a_0)$.

APPENDIX VIII

by J. M. Bloom

Deformation Plasticity Failure Assessment Diagram (Participant 6)

The method used in the ASTM E24.06.02 Predictive Round Robin on Fracture was based on the Central Electricity Generating Board (CEGB) of the United Kingdom failure assessment approach, referred to in the United Kingdom as R-6. This method is an engineering approach to the elastic-plastic fracture mechanics assessment of structural components.

The use of the R-6 diagram allows a straightforward prediction of maximum load which a given structure can safely withstand. In addition, for a given load and either a postulated or actual defect size, the margin of safety of the structure can be determined directly from the diagram. The original concept of the R-6 approach came from the CEGB Dowling and Townley two-criteria approach [*13*] which states that a structure will fail by either of two mechanisms: brittle fracture or plastic collapse. These two mechanisms are connected by a transition curve based on the strip yield model [*14*]. Harrison et al of CEGB [*15*] reformulated the two-criteria approach into what is known today as R-6.

Reference *25* illustrates and explains the use of the R-6 failure assessment diagram (FAD) in predicting the ductile tearing behavior of compact fracture test specimens of HY130 steel, A533B steel, and 7075-T651 aluminum. While it is known that the R-6 assessment curve is based upon the elastic-perfectly plastic Dugdale strip yield model, Ref *25* used a simple approach to account for strain-hardening effects. This approach was to take the average of the yield stress and ultimate tensile strength of the material as the flow stress. While this simple approach gave fairly accurate predictions of maximum load values for the materials in Ref *25*, it was found that prediction of maximum load for both 2024-T351 aluminum alloy and 304 stainless steel compact (baseline) specimens was in considerable error. An alternative failure assessment curve was then developed [*16*] which takes into account the strain-hardening behavior of the 2024-T351 aluminum alloy and 304 stainless steel materials.

The derivation of the new assessment curve, which includes strain-hardening effects, is based upon the hypothesis that the FAD can be represented as a universal failure curve that is normalized with respect to the linear elastic fracture mechanics (LEFM) behavior of the specific structural configuration at one extreme, and the limit load of the structure at the other extreme. In terms of the J-integral, the equation of the failure assessment curve can be written as

$$J_{\text{elastic}}/J = f(S/S_L) \qquad (41)$$

where

$f(\ ) = $ functional relationship,
$S = $ applied stress on the structure or specimen, and
$S_L = $ limit stress.

Now, when $J \leq J_{\text{material}}$, the structure can be considered safe. For crack initiation J_{material} corresponds to J_{Ic}, and for ductile tearing J_{material} would correspond to the appropriate value of the material resistance, J_R. The function $J_{elastic}/J$ can take on various forms, depending on what crack configuration is to be analyzed and on what plasticity model is to be used.

The DPFAD used to account for strain hardening effects in the round robin was derived from an expression for $J_{elastic}/J$ based upon the power-law hardening theory of deformation by Shih and Hutchinson [26] for an infinitely wide middle-crack specimen under plane-stress conditions and the estimation scheme of Shih and Kumar. The derived expression is a function of the material stress-strain curve. It must be emphasized that this is identical to the current R-6 assessment curve approach; that is, the resulting assessment curve is still based upon an infinitely wide middle-crack panel, but with normalization of the end points of the curve by the LEFM behavior of the structure in terms of K_r, as well as the limit load behavior of the structure in terms of S_r. Both S_r and K_r are the original assessment coordinates of the R-6 failure assessment diagram [15].

To predict the failure loads on the round-robin specimens, the coordinate points (S_r, K_r) of the FAD were determined at a fixed load level such that the coordinate points would remain inside the FAD. The expressions used to calculate these points were

$$K_r^2 = J_{le}(a + \Delta a_p)/J_R(\Delta a_p) \qquad (42)$$

and

$$S_r = S/S_{limit} \qquad (43)$$

where

$J_{le}(a + \Delta a_p) = $ J value of the elastically calculated stress-intensity factor for the desired specimen with a crack "$a + \Delta a_p$,"

$J_R = $ experimentally given J-resistance curve for the particular material and size specimen given in the round robin,

$S = $ applied stress (here a predetermined value such that the coordinates lie within the FAD), and

$S_L = $ plastic collapse stress of the specimen to be predicted.

Each increment of crack growth produces a point on the FAD, and these points are scaled up to produce the actual load behavior of the specimen as a function of crack growth. This procedure is explained in more detail in Ref 16.

Figure 49 illustrates the use of the DPFAD with stable crack growth. The position of the assessment points (symbols) relative to the failure assessment curve determines how close the cracked specimen is to initiation of ductile tearing, as shown by point L_1. This point was calculated using the initial crack length and a predetermined load such that the point lies within the FAD. The other points were calculated at this predetermined load for various amounts of crack extension. Because S_r and K_r are directly proportional to applied load, the applied load can be scaled until the point L_1 moves radially from the origin of the diagram to the point L_i. L_i is the initiation point of ductile tearing. The dashed curve shows the translation of all assessment points. After initiation, the locus of (S_r, K_r) points will follow the FAD, as indicated by the arrow, between L_i and the maximum load point, L_m. The dash-dot curve shows the translation of all assessment points at maximum load. Thus, the FAD defines the region between stable and unstable crack growth. For displacement-controlled specimens, the locus of (S_r, K_r) points will follow the FAD. For load-controlled specimens, the locus of points will go outside the FAD, as indicated by the dash-dot curve, after the maximum load point has been reached, indicating that the specimen has become unstable.

The DPFAD approach is similar to an effective J_R-curve method with a limit-load condition except for the approximation to universalize the effects of specimen and crack geometry. Subsequent work has shown that these effects may be significant [27].

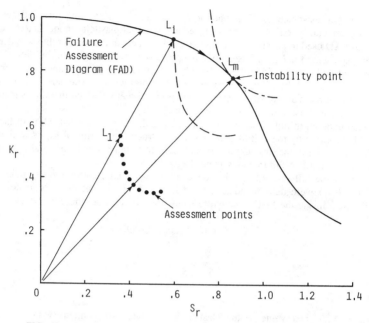

FIG. 49—*Strain-hardening failure-assessment diagram for a given material.*

Because 7075-T651 aluminum had little strain hardening, the R-6 failure assessment curve was used to predict failure loads. This is consistent with the conclusions of Ref *16.* For 7075-T651, three physical J_R-curves (K_R^2/E) were used. These curves were based on the three compact specimen sizes ($W = 51$, 102, and 203 mm) and they were used to predict failure loads on the corresponding specimen sizes. For the other cracked specimens, the J_R-curve for the 102-mm-wide compact specimen was used for the 125-mm-wide MT specimens and the J_R-curve for the 203-mm-wide compact specimen was used for the 250-mm-wide MT and THT specimens. The original predictions for the 7075-T651 aluminum MT and the THT specimens used incorrect K_R-curves. The corrected calculations changed the standard errors from 0.16 to 0.12 for the MT specimens and from 0.22 to 0.11 for the THT specimens. These standard errors are now consistent with Participant 15.

For 2024-T351 aluminum alloy and 304 stainless steel, separate FAD's were generated based upon a power-law hardening stress-strain curve fit to the true stress-true plastic strain data of the respective materials. The J_R-curve used for the 2024-T351 material was an upper-bound fit to the data on both the 102- and 203-mm-wide compact specimens. This derived curve was used in the prediction of all 2024-T351 specimens regardless of specimen width.

For 304 stainless steel, two distinct J_R-curves were used: one for the 102-mm-wide compact specimens and the other for the 203-mm-wide compact specimens. For the 51-mm-wide compact specimens, the J_R-curve for the 102-mm-wide specimen was used. For the other cracked specimens, the J_R-curve for the 102-mm-wide compact specimen was used for the 125-mm-wide MT specimens and the J_R-curve for the 203-mm-wide compact specimen was used for the 250-mm-wide specimens.

The appropriate limit-load expressions were taken from the open literature for both the compact and MT specimens, while the THT specimen limit load was developed on the basis of a middle-crack plate in tension with an effective crack length of the "hole-diameter plus crack length."

APPENDIX IX

by G. E. Bockrath, J. B. Glassco, and D. O'Neal

Theory of Ductile Fracture (Participants 7–9)

Two fracture-analysis methods were used in the predictive round robin. The critical elastic stress-intensity-factor approach accounting for width effects was used for the compact specimens, and the Theory of Ductile Fracture was used for MT and THT specimens. The critical stress-intensity-factor approach is like that described in Appendix III. Some details of the Theory of Ductile Fracture are described herein.

The Theory of Ductile Fracture [*17*] has been developed over a period of about 17 years. During that period, the theory had been extended to include several crack configurations and to account for specimen width effects. The equation for the critical crack length at the ultimate tensile strength using only data from a full-range stress-strain curve had been developed. Also, the critical crack length at the elastic limit stress could also be calculated. From fracture test data, it was shown that these two points were connected by a straight line on a log failure stress against log crack length plot. An equation for the maximum thickness where plane-stress conditions exist had also been determined. The theory had been developed for both through cracks and surface cracks, for various temperatures, and for various materials. See Ref *17* for further details.

The original predictions submitted to the predictive round robin were based on the full stress-strain curves provided (Figs. 3, 16, and 22) as baseline tensile data. For the 2024-T351 material, the failure load predictions on the MT specimens were good (within 10% of experimental failure loads). But the failure load predictions on the 7075-T651 aluminum alloy and 304 stainless steel MT specimens were about 50% higher than experimental failure loads. Because of the large errors on these materials, new stress-strain curves from specimens cut from the same stock of material were conducted at California State University. The new stress-strain curves were from 2.54-mm-thick tension specimens cut from the center of the 12.7-mm-thick plates. The new curves were appreciably different from those provided as baseline tensile data (see Fig. 50 for 7075-T651). Since the new curves gave different failure strains, it could be concluded that there is some difference in the fracture properties between the material at the center of the plate and the material at the surface. (The stress-strain curves for the 304 stainless steel specimens, not shown, gave σ_{ys} and σ_u within a few percent. The failure strains, however, differed by 20%.)

The new failure load predictions (not shown in this report) on the 7075-T651 and 304 MT specimens using the new stress-strain curves were in good agreement with the experimental failure loads. The error on the 7075-T651 specimens went from 64% to less than 8% and that on the 304 specimens went from 45% to, again, less than 8%. Part of the improvement for the 304 stainless steel was due to choosing a higher yield point on the stress-strain curve to fit the Ramberg-Osgood [*7*] equation. This gave a much better fit over the stress-strain curve from the elastic limit to the ultimate tensile strength.

For the THT specimens, the new stress-strain curves and a reevaluation of the effects of stable crack growth gave much better results (errors reduced from 75% to 100% to, again, less than 8% on the 7075-T651 aluminum alloy material).

APPENDIX X

by R. deWit

K_R-Curve with Dugdale Model (Participant 10)

The fracture-analysis method used was the K_R-curve with the Dugdale model. The K_R-curve considered here is the effective K_R plotted against the physical crack length, Δa_p.

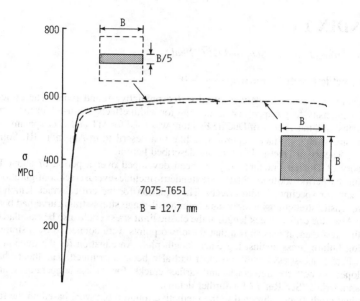

3257 (ASTM) STP-896, Folder 3c, las (1x AAs), Gal. , TTI6 3257$$$$3C, 10-16-85 10-43-

The Dugdale model is used to calculate the effective K_R. The method consists of two steps: (1) Determine analytic expressions for the K_R-curves, and (2) determine the failure load which makes the "crack-driving-force" curve tangent to the K_R-curve.

K_R-Curve

The analytic expressions for the K_R-curves were determined from the baseline compact specimen fracture data. Several expressions discussed by Orange [28] were tried for the 7075-T651 aluminum alloy specimens. The expression that gave the smallest residual error was

$$K_R{}^2 = \frac{c_1{}^2 \, \Delta a_p}{c_2 + \Delta a_p} \qquad (44)$$

where $\Delta a_p = a - a_0$. (Orange showed that this expression gives an equivalent K_R-curve which corresponds to Newman's Two-Parameter Fracture Criterion [11]). Equation 44 was fitted to the physical crack-extension data in Tables 17 and 18 by using a least-squares nonlinear regression analysis. First, the effective K_R values were computed, using Eq 2 with $\lambda = a_e/W$, where a_e/W is the average of the two columns for V_o and V_{LL} in the tables. Next, the physical crack length, a, was also determined from the average between the lengths obtained from unloading compliance at V_o and V_{LL}. The values of the parameters c_1, c_2, and a_0 that were determined to best fit the experimental data are given in Table 21.

The data from Table 19 (304 stainless steel) were not fitted because they did not seem to follow a typical K_R-curve. Especially disconcerting was the fact that during the loading of the steel specimens the relative physical crack lengths (a/W) initially were considerably

Due to an erroneous printing error, Figure 50, page 80 appears incorrectly in this volume of STP 896. Please use the figure and caption below as the correct version for Figure 50. We regret any inconvenience this may have caused.

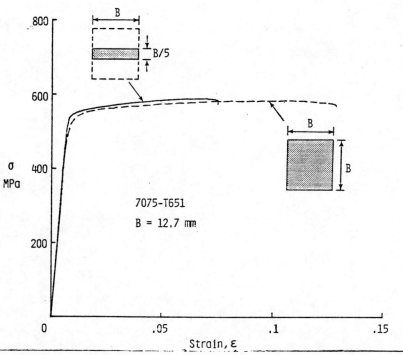

FIG. 50—*Comparison of stress-strain curves from full thickness and one-fifth thickness specimens for 7075-T651 aluminum alloy.*

TABLE 21—*Curve parameters fitted to compact specimen fracture data.*

Material	W, mm	c_1, MN/m$^{3/2}$	c_2, mm	a_0, mm
7075-T651	203	57.8	3.90	102.1
	102	52.6	2.67	51.4
	51	48.4	2.11	25.4
2024-T351	203	167.0	5.24	102.3
	102	142.9	2.49	50.6
	51	126.5	1.99	26.3

less than a_0/W. This is the reason why the steel specimens were not analyzed. (The visual crack length data were not supplied in the original round robin.)

The K_R-curve parameters were dependent upon specimen width. Therefore, the values of c_1 and c_2 for the MT and THT specimens were obtained by interpolation. A parabola was fitted through the three values of c_1 and c_2, separately, for each material. The values of c_1 and c_2 obtained from this parabola at $W/2$ were then used for the MT specimens. The factor of 2 was used because the MT specimen was regarded (crudely) as two compact specimens facing each other. The results of the interpolation are shown in Table 22. Because the THT specimen resembles the MT specimen, the values of c_1 and c_2 for $W = 254$ mm were also used for that specimen configuration.

Crack-Driving-Force Curve

The D-BCS-HSW model was used for the crack-driving-force curve. This model is based on the plastic strip-yield model of Dugdale [*14*], elaborated by Bilby et al [*29*], and extended to finite configurations by Heald et al [*18*]. According to this model, the crack-driving force is given by the effective stress-intensity factor [*30*]

$$K_e = \sigma_o FN \left[\frac{8a}{\pi} \ell n \sec \left(\frac{\pi S}{2N\sigma_o} \right) \right]^{1/2} \tag{45}$$

where

F = boundary-correction factor,
N = nominal stress conversion factor,
S = gross applied stress,
a = physical crack length, and
σ_o = effective flow strength.

For small applied stress, Eq 45 reduces to

$$K_e = S\sqrt{\pi a}\, F \tag{46}$$

TABLE 22—*Interpolated parameters for K_R-curves for MT and THT specimens.*

Material	W, mm	c_1, MN/m$^{3/2}$	c_2, mm
7075-T651	127	49.6	2.24
	254	54.3	2.97
2024-T351	127	131.9	2.06
	254	150.1	2.95

Boundary-Correction Factor

The boundary-correction factor depends upon specimen type. For the compact specimen, F is given by

$$F = \frac{(2 + \lambda)(0.866 + 4.64\lambda - 13.32\lambda^2 + 14.72\lambda^3 - 5.6\lambda^4)}{(\pi\lambda)^{1/2}(1 - \lambda)^{3/2}} \tag{47}$$

where $\lambda = a/W$. For the MT specimen, F is given by

$$F = (\sec \pi\lambda)^{1/2} \tag{48}$$

For the THT specimen, F is given by Eq 21.

Conversion Factor on Nominal Stress

The conversion factor on nominal stress, N, also depends upon specimen type and loading. It represents the ratio of applied stress to nominal stress at the crack tip. The nominal stress at the crack tip is a "strength of materials" calculation of local stress at the crack tip without considering the crack-tip stress concentration. Expressions of the nominal stress for the compact and MT specimen are given in ASTM Standard Terminology (E 616).

For the compact specimen, N is given by

$$N = \frac{(1 - \lambda)^2}{2(2 + \lambda)} \tag{49}$$

For the MT specimen, it is given by

$$N = 1 - 2\lambda \tag{50}$$

For the THT specimen, it is given by

$$N = 1 - (D + a)/W \tag{51}$$

where a is measured from the edge of the small hole (see Fig. 1c).

Effective Flow Strength

The effective flow strength is the nominal stress required to produce a fully plastic region on the net section and was given by

$$\sigma_o = \beta(\sigma_{ys} + \sigma_u)/2 \tag{52}$$

The values of the yield stress and ultimate tensile strength are given in Table 4. The factor β is a constraint factor which depends on the specimen type. From Newman [12], $\beta = 1.62$ for the compact specimen and $\beta = 1$ for the MT specimen. The value of β was assumed to be unity for the THT specimen.

Determination of Failure Load

According to the K_R-curve method, the failure load (or applied stress) is determined by finding the load (or applied stress) which makes the "crack-driving-force" curve tangent to the K_R-curve as a function of crack extension. The K_R-curve was given by Eq 44 with the appropriate values of c_1 and c_2. The value of applied stress, S, in Eq 45 was adjusted until the crack-driving-force curve was tangent to the K_R-curve. The failure load P was given by the applied stress times the gross section area (WB).

APPENDIX XI

by T. W. Orange

Effective K_R-Curve Estimated from Failure Load Data (Participant 11)

The fracture-analysis method that was used is described in Ref *19*. Briefly, the effective K_R-curve for each material was estimated by using only the residual-strength data provided on the baseline compact specimens. Then, the failure loads for other cracked specimens were calculated from the estimated K_R-curves using conventional instability analysis [*31*]. Details are as follows.

The first step requires that the residual-strength (compact specimen) data be differentiated numerically. To do this, the equation

$$P_f/B = A_1 a_0 + A_2 a_0^2 + A_3 a_0^3 \tag{53}$$

was fitted to the residual-strength data for each material. The first derivative of P_f/B with respect to a_0 is given by

$$\frac{1}{B}\left(\frac{dP_f}{da_0}\right) = A_1 + 2A_2 a_0 + 3A_3 a_0^2 \tag{54}$$

For each material, the polynomials (Eqs 53 and 54) were evaluated at $a_0 = 12.5, 25, 37.5, 50, 75,$ and 100 mm. These values were substituted into Eq 76 of Ref *19*, which becomes

$$2\left(\frac{1}{P_f}\frac{dP_f}{da_0} - \frac{1}{a_0}\right) + \frac{1 - 2\alpha(\Delta a_e/a_0)}{a_0 + \Delta a_e} = 0 \tag{55}$$

where

$$\alpha = \frac{\lambda}{Y}\frac{dY}{d\lambda} \tag{56}$$

$$\lambda = (a_0 + \Delta a_e)/W \tag{57}$$

$$Y = \frac{(2 + \lambda)}{\lambda^{1/2}(1 - \lambda)^{3/2}}(0.886 + 4.64\lambda - 13.32\lambda^2 + 14.72\lambda^3 - 5.6\lambda^4) \tag{58}$$

and Δa_e was determined from Eq 55 for each value of a_0 using an iterative procedure [*19*]. The function Y is the stress-intensity coefficient for the compact specimen. The value of K_R was then calculated from the failure load (P_f) and effective crack length ($a_0 + \Delta a_e$). These six values of K_R and Δa_e gave the estimated effective K_R curve for each material. So that instability calculations could be done numerically, equations were fit to each K_R curve and these are given as

MATERIAL	K_R-CURVE EQUATION	
7075-T651	$K_R^2 = 3388 \ (1 - e^{-0.1378\Delta a_e})$ for $\Delta a_e < 15$ mm	(59)
2024-T351	$K_R^2 = 1330\Delta a_e^{0.8738}$ for $\Delta a_e < 18$ mm	(60)
304	$K_R^2 = 4396\Delta a_e^{0.8026}$ for $\Delta a_e < 15$ mm	(61)

where Δa_e is in millimetres and K_R is in MN/m$^{3/2}$. The choice of the forms for these equations was based on intuition and experience. For the aluminum alloys, these equations were in good agreement with the measured effective K_R-curves. A comparison between the measured (symbols) and estimated (curve) effective K_R-curves for 7075-T651 aluminum alloy is shown in Fig. 51. The estimated K_R-curve from Eq 59 is restricted to Δa_e less than 15 mm (solid curve). For the stainless steel, the estimated and measured K_R-curves have similar magnitudes but considerably different shapes. If one computes nominal K values for the baseline stainless steel specimens using maximum load and original crack length, it is obvious that the ASTM E 561 ligament size requirement was grossly violated for all specimen sizes. In retrospect, perhaps the predictions for the stainless steel should not have been attempted at all. In any event, Eq 61 represents a pseudo-K_R-curve whose significance (if any) is not obvious.

Numerical calculation of K_R-curve instability followed standard procedures as given in Ref *31*. All calculations were done on a programmable desk calculator. The well-known secant K expression was used for the MT specimens. Because of calculator storage limitations, low-order spline functions were used for the compact and THT specimens. As discussed in Ref *19,* this method has the same limitation as a standard K_R-curve test

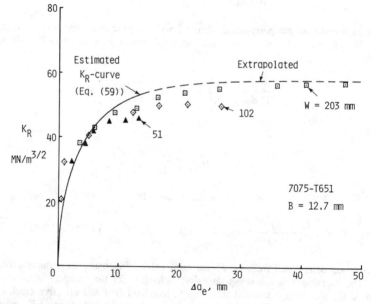

FIG. 51—*Comparison of K_R-curve obtained from residual-strength data and measured on 7075-T651 aluminum alloy compact specimens.*

conducted under load control; namely, the curve can be measured (or estimated) only up to the point at which the crack becomes unstable, hence the Δa_e limits given earlier for the K_R equations. Instability calculations which required significant extrapolation were not reported. Also, the calculated instability conditions were compared with the ASTM E 561 size requirements. If these requirements were exceeded, the calculated failure load was not reported and it was assumed that a limit-load failure would occur. For the MT and THT specimens, the net-section stress based on physical crack length must be less than the yield stress of the material. For the compact specimen, the remaining uncracked ligament $(W - a)$ must be greater than $4/\pi(K_{max}/\sigma_{ys})^2$ where K_{max} is the maximum K level in a test.

APPENDIX XII

by G. A. Vroman

Effective K_R-Curve and Limit-Load Criteria (Participant 12)

The fracture-analysis method used was the effective K_R-curve method as described in ASTM E 561 except that the net-section stress requirement, $S_n < \sigma_{ys}$, was replaced by $S_n < \sigma_o$. The flow stress (σ_o) was assumed to be $(\sigma_{ys} + \sigma_u)/2$.

The effective K_R-curve for 7075-T651 and 2024-T351 aluminum alloy was obtained from the baseline compact specimens. The method was not applied to the 304 stainless steel specimens.

Predicted failure loads were made on only the THT specimens. The "crack-driving force" curve for the THT specimen was calculated from the stress-intensity factor equation (Eq 20). Failure load was determined by finding the value of load such that the crack-driving force curve would be tangent to the effective K_R-curve. If the net-section stress exceeded σ_o, before tangency was met, then the net-section failure stress at the current effective crack length was assumed to be σ_o. Failure load was $P = \sigma_o A_{net}$.

APPENDIX XIII

by R. J. Allen

Effective K_R-Curve and Limit-Load Criteria (Participant 13)

The fracture-analysis method used for the 7075-T651 and 2024-T351 aluminum alloy specimens was the effective K_R-curve approach with a limit-load condition. A plastic-collapse (or limit-load) analysis was used for the 304 stainless steel specimens.

Steel Specimens

All the 304 stainless steel specimens were assumed to fail by plastic collapse (or limit load) with only minimal stable crack growth (or tearing). The limit loads were calculated from the following relations.

$$\text{Compact:} \qquad P_L = 0.6\sigma_o B \frac{(W - a)^2}{W + a} \qquad (62)$$

$$\text{MT:} \qquad P_L = \sigma_o B(W - 2a) \qquad (63)$$

$$\text{THT:} \qquad P_L = \sigma_o WB(1.03 - 2.24a) \qquad (64)$$

In the last equation, crack length a is given in metres. The crack length was assumed to be the initial crack length, a_0. The effective flow stress, σ_o, was the geometric mean of the yield stress and the ultimate tensile strength, that is, $\sigma_o = (\sigma_{ys}\sigma_u)^{1/2}$.

The methods used to calculate the upper and lower bounds for the plastic-collapse loads on the THT specimens are illustrated in Fig. 52. The upper and lower bounds were obtained from the following theorems:

1. If a deformation pattern can be invented which is compatible with the displacement boundary conditions and involves no volume change, then the work which would be done by the external load in a small displacement (at incipient collapse) is less than or equal to the plastic work done on the assumed shear surfaces. This gives an upper bound on the limit load (see Fig. 52a).
2. Consider any stress distribution which could be equilibrated by a load of the type applied. The maximum value of load consistent with stresses everywhere at or below yield is a lower bound on the limit load (see Fig. 52b). The ratio σ_A/σ_B was determined from the zero-net-moment requirement, and the longest of σ_A/σ_B was set equal to σ_o.

The normalized limit load ($P_L/\sigma_o BW$) is plotted against crack length measured from the edge of the small hole in Fig. 53. Values given by Eq 64 lie roughly halfway between the upper-bound (solid curve) and lower-bound (dashed curve) estimates. The upper- and lower-bound estimates were separated by about a factor of 2. Thus, the possibility of error in predicting failure loads on the THT specimens is substantial. (In fact, the test results indicate that something closer to the lower bound would have been appropriate.)

For the compact specimen, the upper and lower limit loads were

$$0.5\sigma_o B \frac{(W - a)^2}{W + a} \leq P_L \leq 0.69\sigma_o B \frac{(W - a)^2}{W + a} \qquad (65)$$

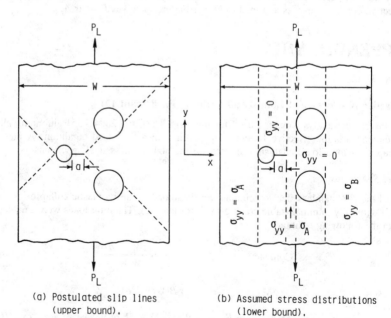

(a) Postulated slip lines
(upper bound).

(b) Assumed stress distributions
(lower bound).

FIG. 52—*Upper and lower bounds on plastic-collapse load for three-hole-crack tension specimen.*

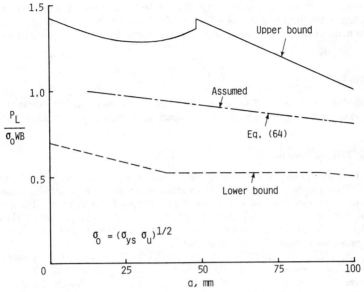

FIG. 53—*Plastic-collapse load calculations for three-hole-crack tension specimen.*

Here only the bending component was considered. An average between the upper and lower bounds (Eq 62) was used.

For the MT specimen, the upper and lower bounds were identical and were given by Eq 63.

Aluminum Specimens

For the 7075-T651 and 2024-T351 specimens, the progress of crack extension was predicted from the effective K_R-curve determined from the baseline compact specimens. The K_R-curves are shown in Figs. 6 and 18 for 7075-T651 and 2024-T351 materials, respectively. The result for the large compact specimen ($W = 203$ mm) was used for the 7075-T651 material. The "crack-driving force" curves were determined from the stress-intensity factor solution for the respective specimens. Failure load was determined by the tangent point between the crack-driving-force curve and the effective K_R-curve, unless the specimen failed by plastic collapse. The plastic-collapse load was calculated by substituting the current effective crack length, a_e, into Eq 62 or Eq 63 or Eq 64.

APPENDIX XIV

by D. E. McCabe

Effective K_R-Curve and Limit-Load Criteria (Participant 14)

Compact Specimens

The procedure applied to predict instability loads on compact specimens made of 7075-T651 and 2024-T351 materials was the effective K_R-curve practice as described in ASTM Method E 561.

A limit-load criterion was used on the 304 stainless steel compact specimens. A brief description of the limit-load criterion follows. From the baseline compact specimens, an average nominal net-section failure stress (S_n) was computed for each specimen width as

$$S_n = \frac{2P_f}{B(W - a_0)^2}(2W + a_0) \qquad (66)$$

The value of S_n was assumed to be constant at failure for each specimen width, and the failure loads were computed for the other compact specimens ($a_0/W = 0.3$ and 0.7) using Eq 66.

MT and THT Specimens

For these specimens, the effective K_R-curve practice as described in ASTM E 561 was applied. The K_R-curve representation used was K_R against Δa_e. The "crack-driving-force" curve was calculated from the appropriate stress-intensity factor solution. This approach is usable only up to limit load. If tangency between crack-driving-force curve and the K_R-curve has not developed at net-section stress equal to an effective flow stress (σ_0), then failure was calculated by the limit-load criterion.

Only the 7075-T651 material failed according to the K_R concept. To demonstrate the use of the K_R-curve concept, the failure load on the THT specimen with $a_0 = 25.5$ mm will be predicted. Figure 54 shows K and K_R plotted against crack length. The solid curve shows the K_R-curve for the 7075-T651 material obtained from the baseline compact specimens (see Fig. 6). The dashed curve is an estimated extrapolation of the K_R-curve. The dash-dot curves are the crack-driving-force curves calculated from Eq 20 with various applied loads. At $P = 200$ kN, the crack-drive curve intersects the K_R-curve at Point A and the crack has extended about 1.5 mm. At $P = 450$ kN, the crack-drive curve intersects

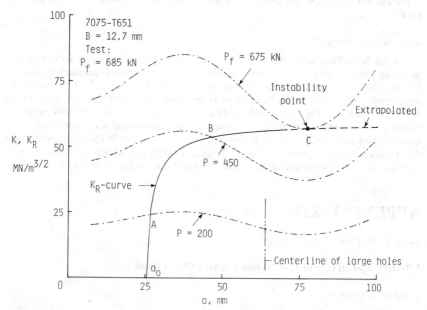

FIG. 54—*Use of K_R-curve concept to predict failure load on three-hole-crack tension specimen made of 7075-T651 aluminum alloy.*

the K_R-curve at Point B and the crack has extended about 20 mm. The predicted failure load was 675 kN. At this load, the crack-drive curve is tangent to the K_R-curve at Point C, the instability point. For any crack extension beyond Point C, the crack-drive curve is higher than the material resistance indicating instability. The instability point occurs about 15 mm past the centerline of the large holes (dash-double-dot line) and at the minimum point on the crack-drive curve. This corresponds quite well with the experimental results shown in Figs. 13 and 14a.

The 2024-T351 aluminum alloy and 304 stainless steel specimens failed in limit-load condition. For the 2024-T351 material, the limit-load condition was given by the net-section stress equal to σ_o ($= 335$ MPa). For the 304 stainless steel, two flow stresses were used. A flow stress of 470 MPa was used for the 102-mm-wide compact specimens and the 127-mm-wide MT specimens. A flow stress of 390 MPa was used for the 203-mm-wide compact specimens and the 254-mm-wide MT and THT specimens.

APPENDIX XV

by J. D. Landes

Effective K_R-Curve, Ligament Analysis, and Limit-Load Criteria (Participant 15)

The methods used in the predictive round robin are presented here in schematic form. The methods were not always strictly applied; some judgment (or guess) was made.

Aluminum Alloy 7075-T651

Compact Specimens—The method used for the compact specimens is a graphical interpolation (or extrapolation) method using K_R and Δa_e. The method is a simple application of effective K_R-curve procedure which avoids the laborious procedure of constructing K "crack-driving-force" curves for different loads to determine the tangent point. The tangent (or instability) point is estimated by using the following procedure.

From the baseline compact specimen data, the tangent point is determined for the three specimen sizes. For example, for the 203-mm compact specimen (Table 17) the maximum load is 24.05 kN. At this load, K_R is 46.5 MN/m$^{3/2}$ and $\Delta a_e = 10.4$ mm using the results in the $(a_e/W)_o$ column in Table 17. The values of K_R and Δa_e are then plotted as a function of ligament size, $W - a_0$, for the three specimen sizes as illustrated in Fig. 55. A smooth curve is drawn through these points. To determine the maximum (failure) load on the other compact specimens, the ligament size ($W - a_0$) is calculated and the value of K_R and Δa_e are then determined from Figs. 55a and 55b, respectively, as illustrated by the dashed lines. The effective crack length is $a_e = a_0 + \Delta a_e$ and the failure load is

$$P = \frac{K_R B \sqrt{W}}{F(a_e/W)} \tag{67}$$

where $F(a_e/W)$ is the boundary-correction factor for the compact specimen written in terms of a_e/W.

Middle-Crack Tension Specimens—For these specimens, the point of tangency between the K "crack-driving-force" curve and the effective K_R-curve was estimated by first assuming a straight line for the crack-driving force curve to get a tangency point and then backing down the curve a little to compensate for the roundness in the actual crack-driving force curve. Because the K_R-curve is fairly flat (see Fig. 6), a fair estimate of the tangency point can be made. The K_R-curve for the 203-mm-wide compact specimen was used.

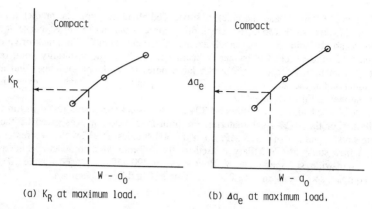

FIG. 55—*Graphical method to determine failure load and crack extension on compact specimens made of 7075-T651 and 2024-T351.*

From the tangency point, the values of K_R and Δa_e were then used to calculate failure load as

$$P = \frac{K_R W B}{\sqrt{\pi a_e} \, F(a_e/W)} \qquad (68)$$

where $a_e = a_0 + \Delta a_e$ and $F(a_e/W)$ is the boundary-correction factor for the MT specimen.

Three-Hole-Crack Tension Specimens—For this configuration, two procedures were used depending upon the initial crack length. For a_0/W ratios less than or equal to 0.3, it was assumed that the minimum point on the K calibration curve (see Fig. 44) at $a/W = 0.3$ would be tangent to this flat part of the K_R-curve (Fig. 6). A value of K_R was picked and the failure load was calculated from Eq 68 where F is the boundary-correction factor for the THT specimen and F was given by Eq 21 with $a_e = 0.3 \, W$.

For initial crack lengths greater than 0.3, the procedure described for the MT specimens was used to estimate the tangent point. Again, from K_R, Δa_e, and the K calibration curve (Eq 20), the failure load was calculated.

Comment on 7075-T651 Aluminum Alloy—For all crack configurations, the LEFM approach with K_R and Δa_e was used. The usual procedure of determining the tangent point between the K_R-curve and the "crack-driving-force" curve was not used. Rather, an estimation procedure was used to determine the tangent point. One observation made on this material was that the K_R-curve determined from compact specimens is too low for the tension-type specimens. The minimum point on the K calibration curve for the THT specimen should reach tangency far out on the flat part of the K_R-curve. Based on the actual failure loads, the K_R-curve should have been about 10% higher than the one provided for the compact specimen.

Aluminum Alloy 2024-T351

Compact Specimens—The same procedure described for the 7075-T651 compact specimens was used for these specimens also.

MT and THT Specimens—For these specimens, a limit-load criterion was used. A flow stress (σ_0) of 415 MPa (or 60 ksi) was chosen, $\sigma_0 = 1.1(\sigma_{ys} + \sigma_u)/2$. The failure load was $P = \sigma_0 A_{net}$ where A_{net} was the net-section area based on the initial crack length.

Comment on 2024-T351 Aluminum Alloy—The effective K_R-curve from the compact

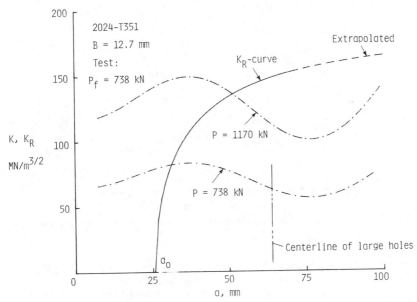

FIG. 56—*Attempt to use K_R-curve concept to predict failure load on three-hole-crack tension specimen made of 2024-T351 aluminum alloy.*

specimens is not appropriate for the tension specimens. As an example, an attempt was made to use the effective K_R-curve concept to predict failure load on the THT specimen with an initial crack length of 25.7 mm. Figure 56 shows K and K_R plotted against crack length, a. The solid curve shows the K_R-curve fitted to the results from the baseline compact specimens (see Fig. 18). The upper portion of the K_R-curve (dash curve) was extrapolated. Using Eq 20 to calculate crack-driving-force curves, two curves were calculated. One curve, the upper dash-dot curve, was calculated using a failure load (1170 kN) based on applying the ultimate tensile strength (σ_u) on the initial net-section area. As can be seen, the upper dash-dot curve is far from being tangent to the compact specimen K_R-curve. The lower dash-dot curve was calculated using the actual failure load (738 kN) on this specimen (see Table 6c). Thus, the "true" K_R-curve for the tension specimen should be much lower than that for the compact specimen. And, consequently, the effective K_R-curve is specimen-type dependent for this material.

Stainless Steel 304

Compact Specimens—The nominal net-section failure stress, S_n, was calculated from the baseline compact specimen data as

$$S_n = \frac{2P}{B(W - a_0)^2} (2W - a_0) \qquad (69)$$

using the maximum (failure) loads from Table 7a. These nominal stresses were plotted against ligament size, $W - a_0$, as illustrated in Fig. 57. It was assumed that there was no physical crack growth at maximum load (or that any effect of physical crack growth was incorporated into Fig. 57).

The failure loads on the other compact specimens were predicted by calculating the

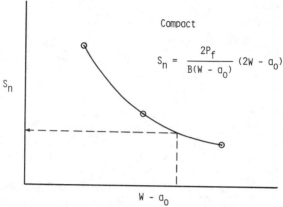

FIG. 57—*Graphical method to determine failure load on compact specimens made of 304 stainless steel.*

ligament size and determining the nominal failure stress (S_n) from Fig. 57, as illustrated by the dashed lines. The failure load was then calculated from Eq 69.

MT and THT Specimens—For these specimens, a limit-load criterion was used. The flow stress (σ_o) was, again, estimated as $1.1(\sigma_{ys} + \sigma_u)/2$. The flow stress was 480 MPa (70 ksi). Failure loads were predicted from $P = \sigma_o A_{net}$ where A_{net} was the net-section area based on initial crack length.

Comment on 304 Stainless Steel—For all crack configurations, a limit-load criterion was used to predict failure loads. This method works quite well if a proper flow stress is chosen. On the basis of the experimental failure loads on the tension specimens, a value of σ_o at about 450 would have been more appropriate. This is about the average between the yield stress and ultimate tensile strength (see Table 4).

APPENDIX XVI

by J. C. Newman, Jr.

Finite-Element Analysis with Critical Crack-Tip-Opening Displacement Criterion (Participant 17)

The fracture-analysis method used was based on a two-dimensional, elastic-plastic (incremental and small strain), finite-element analysis which included the effects of stable crack growth [20,32].

Crack-Growth Criterion

The crack-growth criterion used was a critical crack-tip-opening displacement (δ_c) at a specified distance (d) from the crack tip. The distance d was the element size along the crack line, or, in other words, d is the distance between the first free node and the crack tip in the finite-element model. [The critical CTOD criterion is also equivalent to a critical crack-tip-opening angle (CTOA) criterion, since CTOA $= 2 \tan^{-1} (\delta_c/2d)$.] During incremental loading to failure, whenever the CTOD equaled or exceeded a preset critical value (δ_c), the crack-tip node was released (see Ref 20 for details) and the crack advanced to the next node. This process was repeated until crack growth continued without any increase in load. The use of the CTOD (or CTOA) criterion does require that the absolute size (d) and arrangement of elements in the crack-tip region and along the line of crack extension be the same in all crack configurations considered. Single values of critical CTOD were used in the analysis to model crack initiation, stable crack growth, and instability.

The procedure used to establish the critical δ_c value and mesh size (d) is as follows. The mesh size in the crack-tip region of the large compact specimen ($W = 203$ mm) was systematically reduced until the calculated loads at initiation and at failure were reasonably close to the experimental values using a given value of δ_c. In other words, the mesh size was used as a variable to fit the experimental load against crack length data. After the mesh size d was determined, the final δ_c value was selected so that the mean of the calculated-to-experimental failure load ratio on the various size compact specimens ($W = 51$, 102, and 203 mm) with $a_0/W = 0.5$ was about unity. The critical δ_c value was then used to predict failure loads on other compact, MT, and THT specimens.

A comparison between experimental and calculated load against physical crack extension data on two 7075-T651 aluminum alloy baseline compact specimens is shown in Fig. 58. The symbols show the average experimental crack extension measurements made using visual and unloading compliance (load-line and crack mouth) methods [5]. The respective bars indicate the range and mean of failure (or maximum) load on four or five tests. The bars are placed at the average value of crack extension at maximum load. The solid lines show the calculated crack-growth behavior from the finite-element analysis with $\delta_c = 0.0216$ mm and $d = 0.4$ mm for both specimens. The tests were conducted under displacement-control conditions, whereas the analysis was conducted under load-control conditions. Thus, calculations were not made beyond maximum load. The calculated failure load on the large specimen was about 5% lower than the average experimental failure load. But for the small specimen, the calculated failure load was about 10% higher than the average experimental failure load.

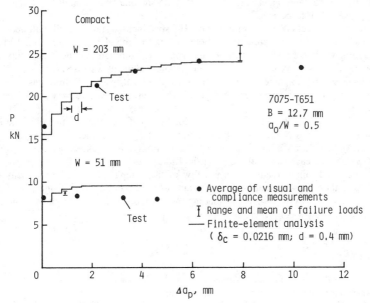

FIG. 58—*Comparison of calculated and experimental crack-growth behavior for 7075-T651 aluminum alloy specimens using finite-element analysis with CTOD concept.*

The critical CTOD values for the three materials were determined from the experimental load against physical crack extension data on the baseline compact specimens. The critical CTOD values are given in Table 11. The distance d (or mesh size) along the crack line was 0.4 mm for all three materials and all specimen sizes considered.

Failure Load Predictions

The failure loads on the compact, MT, and THT specimens were predicted using the finite-element analysis with the δ_c values ($d = 0.4$ mm) determined from the baseline compact specimens. Typical finite-element models for the MT and THT specimens are shown in Ref 20. Predictions were not made on some of the 304 stainless steel specimens because of the large computer cost involved. Further details are given in Ref 20.

APPENDIX XVII

by B. D. Macdonald

Finite-Element Analysis with Critical Crack-Front Singularity Parameter (Participant 18)

The fracture-analysis method used was based on a three-dimensional, elastic-plastic, finite-element analysis with a stationary crack [21,33]. The finite-element analysis was used to calculate the fracture parameter, K_f, from the baseline compact specimen ($W = 203$ mm) data supplied. The fracture parameter K_f is the strength of the singularity for a multilinear hardening material. The fracture parameter was calculated with the ANSYS computer program using Tracey's three-dimensional [34] crack-front element adapted to a multilinear representation of the stress-strain curve. Extrapolating Hutchinson's results for a bilinear hardening material [35], the crack-tip singularity for a multilinear hardening material was taken to be $r^{-1/2}$.

The large compact ($W = 203$ mm) and THT specimens were modeled with about 400 elements (accounting for symmetry). About one half of the elements were in a 13 by 13 by $B/2$-mm volume centered on the crack front. Five through-thickness elements were used along the crack front and one through-thickness element was used outside of the 13 by 13 by $B/2$ volume centered on the crack front. Crack front curvature was estimated from the large compact specimen data. Stable crack growth prior to maximum load was neglected.

Using the initial crack length, the applied load on the finite-element model of the large compact specimen was increased incrementally until the experimental failure load was reached. At the failure load, K_f was computed as a through-the-thickness weighted average. The K_f values for each material are given in Table 12.

The predicted failure loads on only a few of the compact and THT specimens were made using the finite-element analysis. Again, using the initial crack length, the failure load was predicted by incrementally loading the desired specimen configuration until the critical value of K_f was reached.

References

[1] *Plane Strain Crack Toughness Testing of High Strength Metallic Materials, ASTM STP 410*, American Society for Testing and Materials, Philadelphia, 1966.
[2] *Elastic-Plastic Fracture, ASTM STP 668*, American Society for Testing and Materials, Philadelphia, 1979.

[3] *Post-Yield Fracture Mechanics*, D. G. H. Latzko, Ed., Applied Science Publishers Ltd., London, 1979.

[4] *Advances in Elasto-Plastic Fracture Mechanics*, L. H. Larson, Ed., Applied Science Publishers Ltd., London 1980.

[5] McCabe, D. E., "Data Development for ASTM E24.06.02 Round Robin Program on Instability Prediction," NASA CR-159103, National Aeronautics and Space Administration, Aug. 1979.

[6] Clark, G. A., Andrews, W. R., Paris, P. C., and Schmidt, D. W., "Single Specimen Tests for J_{Ic} Determination," *Mechanics of Crack Growth, ASTM STP 590*, American Society for Testing and Materials, Philadelphia, 1976, pp. 27–42.

[7] Ramberg, W. and Osgood, W. R., "Description of Stress-Strain Curves by Three Parameters," NACA TN-902, National Advisory Committee for Aeronautics, 1943.

[8] Srawley, J. E., "Wide Range Stress-Intensity Factor Expressions for ASTM E 399 Standard Fracture Toughness Specimens," *International Journal of Fracture*, Vol. 12, June 1976, p. 475.

[9] Hutchinson, J. W. and Paris, P. C., "Stability Analysis of *J*-Controlled Crack Growth," *Elastic-Plastic Fracture, ASTM STP 668*, 1979, pp. 37–64.

[10] Witt, F. J., "The Equivalent Energy Method: An Engineering Approach to Fracture," *Engineering Fracture Mechanics*, Vol. 14, 1981, pp. 171–187.

[11] Newman, J. C., Jr., "Fracture Analysis of Surface- and Through-Cracked Sheets and Plates," *Engineering Fracture Mechanics Journal*, Vol. 5, 1973, pp. 667–689.

[12] Newman, J. C., Jr., "Fracture Analysis of Various Cracked Configurations in Sheet and Plate Materials," *Properties Related to Fracture Toughness, ASTM STP 605*, American Society for Testing and Materials, Philadelphia, 1976, pp. 104–123.

[13] Dowling, A. R. and Townley, C. H. A., "The Effects of Defects on Structural Failures: A Two-Criteria Approach," *International Journal of Pressure Vessels and Piping*, Vol. 3, 1975.

[14] Dugdale, D. S., "Yielding of Steel Sheets Containing Slits," *Journal of the Mechanics and Physics of Solids*, Vol. 8, 1960, p. 100.

[15] Harrison, R. P., Loosemoore, K., and Milne, I., "Assessment of the Integrity of Structures Containing Defects," CEGB Report R/H/R6, Central Electricity Generating Board, Leatherhead, Surrey, U.K., 1976.

[16] Bloom, J. M., "Prediction of Ductile Tearing Using a Proposed Strain Hardening Failure Assessment Diagram," *International Journal of Fracture*, Vol. 16, 1980, pp. R73–R77.

[17] Bockrath, G. E. and Glassco, J. B., "Theory of Ductile Fracture," California State University Report No. ME 81-400, Long Beach, CA, Nov. 1981.

[18] Heald, P. T., Spink, G. M., and Worthington, P. J., "Post-Yield Fracture Mechanics," *Material Science and Engineering*, Vol. 10, 1972, p. 129.

[19] Orange, T. W., "Method for Estimating Crack-Extension Resistance Curve From Residual Strength Data," NASA TP-1753, National Aeronautics and Space Administration, Nov. 1980.

[20] Newman, J. C., Jr., "Finite-Element Analysis of Initiation, Stable Crack Growth, and Instability Using a Crack-Tip-Opening Displacement Criterion," NASA TM 84564, National Aeronautics and Space Administration, Oct. 1982.

[21] Macdonald, B. D., "Fracture Prediction Based on Plastic Stress Singularity Strength," *Theoretical and Applied Fracture Mechanics*, Vol. 1, 1984, pp. 169–180.

[22] Irwin, G. R., "Analysis of Stresses and Strains Near the End of a Crack Traversing a Plate," *Transactions*, American Society of Mechanical Engineers, *Journal of Applied Mechanics*, 1957.

[23] Raju, I. S. and Newman, J. C., Jr., "Three-Dimensional Finite-Element Analysis of Finite-Thickness Fracture Specimens," NASA TN D-8414, National Aeronautics and Space Administration, May 1977.

[24] Brady, W. G. and Drucker, D. C., "Investigation and Limit Analysis of Net Area in Tension," *Transactions*, American Society of Civil Engineers, Oct. 1953, pp. 1133–1164.

[25] Bloom, J. M., "Prediction of Ductile Tearing of Compact Fracture Specimens Using the R-6 Failure Assessment Diagram," *International Journal of Pressure Vessels and Piping*, Vol. 8, 1980, pp. 215–231.

[26] Shih, C. F. and Hutchinson, J. W., "Fully Plastic Solutions and Large-Scale Yielding for Plane Stress Crack Problems," *Journal of Engineering Materials and Technology*, Vol. 98, No. 4, 1976, p. 289.

[27] Bloom, J. M., "Validation of a Deformation Plasticity Failure Assessment Diagram Approach to Flaw Evaluation," *Elastic-Plastic Fracture: Second Symposium. Volume II—Fracture Re-*

sistance Curves and Engineering Applications, ASTM STP 803, C. F. Shih and J. P. Gudas, Eds., American Society for Testing and Materials, Philadelphia, 1983, pp. II-206-II-238.

[28] Orange, T. W., "A Relation Between Semiempirical Fracture Analyses and R-Curves," NASA TP-1600, National Aeronautics and Space Administration, Jan. 1980.

[29] Bilby, B. A., Cottrell, A. B., and Swinden, K. H., "The Spread of Plastic Yielding From a Notch," *Proceedings of the Royal Society of London*, A272, 1963, p. 304.

[30] deWit, R., "A Review of Generalized Failure Criteria Based on the Plastic Yield Strip Model," *Fracture Mechanics: 14th Symposium, ASTM STP 791*, American Society for Testing and Materials, Philadelphia, Vol. I, 1983, pp. I24-I50.

[31] *Fracture Toughness Evaluation by R-Curve Methods, ASTM STP 527*, American Society for Testing and Materials, Philadelphia, 1973.

[32] Newman, J. C., Jr., "Finite-Element Analysis of Crack Growth Under Monotonic and Cyclic Loading," *Cyclic Stress-Strain and Plastic Deformation Aspects of Fatigue Crack Growth, ASTM STP 637*, American Society for Testing and Materials, Philadelphia, 1977, pp. 56–80.

[33] Macdonald, B. D., "Fracture Prediction Based on Plastic Stress Singularity Strength," *Journal of Theoretical and Applied Fracture Mechanics*, Vol. 1, No. 2, 1984, pp. 169–180.

[34] Tracey, D. C., "Finite Elements for Three-Dimensional Elastic Crack Analysis," *Nuclear Engineering and Design*, Vol. 26, 1974, pp. 282–290.

[35] Hutchinson, J. W., "Singular Behavior at the End of a Tensile Crack in a Hardening Material," *Journal of the Mechanics and Physics of Solids*, Vol. 16, 1968, pp. 13–31.

Elastic-Plastic Fracture
Mechanics Methodology

D. E. McCabe[1] and K. H. Schwalbe[2]

Prediction of Instability Using the K_R-Curve Approach

REFERENCE: McCabe, D. E. and Schwalbe, K. H., **"Prediction of Instability Using the K_R-Curve Approach,"** *Elastic-Plastic Fracture Mechanics Technology, ASTM STP 896,* J. C. Newman, Jr., and F. J. Loss, Eds., American Society for Testing and Materials, Philadelphia, 1985, pp. 99–113.

ABSTRACT: An elastic-plastic R-curve instability prediction method is reported. The method is prepared in a format such that the computational steps outlined can be used as a guide to make instability predictions. Three example problems are given.

KEY WORDS: R-curve, test procedure, instability, fracture mechanics, compliance, plastic zone, ASTM Standard Method E 561, elastic-plastic

This is a predictive practice that is based on a computational methodology outlined in the ASTM Recommended Practice for R-Curve Determination (E 561-81). The E 561 practice had been developed for use on ultra-high-strength sheet materials, but in the present case the usage is liberalized to handle the more ductile structural grades of materials. Under this extension of usage, it is recommended that the determination of plastic zone effects be limited to the experimental compliance approach. As before, the predictive capability is restricted to those cases where the specimens or components are stressed below net section yield.

Instability predictions can be made for any configuration for which a linear elastic K_I analysis exists. The predominant cases handled are for the more simple condition of load control where attainment of maximum load results in an instability event. If, on the other hand, the overall elastic body compliance behavior of the component is known, crack instability for displacement-limited loading conditions can also be predicted.

Significance

The approach taken is that a modified linear elastic methodology can be extended to handle elastic-plastic crack-tip field conditions. The mechanism is the

[1]Senior engineer, Westinghouse Electric R&D, Pittsburgh, PA 15235.
[2]Head, Fracture Mechanics Group, Research Center Geesthacht, Geesthacht, West Germany.

redefinition of a crack size, larger than the physically existing crack, that can be used in a linear elastic K_I expression to develop a plasticity-corrected stress-intensity factor, K_R. K_R-values are equivalent to values of deformation theory J [1]. The appropriate "effective crack size" is determined by compliance methods and the equivalence between K_R and J_R is completely maintained up to ligament yield load in bend configurations and to within 90% of net section yield load in tension configurations [2]. Repetition in usage of this practice, testing various specimen sizes and initial crack sizes, will lead to a realization that material flow property characteristics can be completely incorporated into the R-curve by plotting K_R versus the effective crack growth, Δa_e [3]. Hence, to predict instability, crack drive can be calculated using a pseudo linear-elastic R-curve methodology. The elimination of plastic deformation analysis from crack drive (necessary with J-Δa_p R-curve analysis) then provides a much simplified method for predicting instability. This is of particular value for complex geometries or stress conditions or both for which elastic-plastic J-solutions may not presently exist. Cases of local stress field excursions, perhaps caused by residual stresses, reinforcement straps, or neighboring holes are handled more easily. Again, these analyses require only the knowledge of the elastic K_I stress field behavior of the components.

Definitions

Instability: The condition under which a crack can propagate without added force in load-controlled cases or without added overall displacement under displacement-controlled conditions. Cleavage instability is not necessarily covered by this analysis.

Effective Crack Size: A crack size that includes an initial crack dimension plus stable crack propagation and plastic zone contribution. In this practice, the effective crack size dimension is a compliance equivalent value obtained from secants drawn to the test record [4].

Compliance: The ratio of elastic displacement, $2v$, to load, P, at a chosen location on the specimen or component. The compliance behavior, $(2v)/P$, can be normalized with material thickness, B, and elastic modulus, E, to cover variables of size and material, $(EB2v/P)$.

Plastic Zone: A linear dimension which is added to the physical crack size to account for crack-tip plastic deformation.

Crack Extension: The common J_R-curve representation is usually plotted in terms of physical crack growth, Δa_p, on the abscissa. The present K_R predictive methodology uses effective crack growth Δa_e on the abscissa. Here physical crack growth is augmented with the plastic zone contribution, and both are automatically accounted for using the prescribed secant in the compliance technique. $\Delta a_e = (a_e - a_0)$ where a_0 is the initial crack size.

Laboratory Test Procedure—ASTM Method E 561

The testing procedure is presented in various sections of E 561. As stated in the scope section, the number of geometries that the methodology applies to is not limited, but the specific background information is given on only three specimen designs. The applicable paragraphs taken from E 561 that are pertinent to the present document are as follows:

1. E 561 Scope
   1.2 Materials that can be tested for R-Curve development are not limited by strength, thickness or toughness, so long as specimens are of sufficient size to remain predominantly elastic throughout the duration of the test.
   1.4 Specimens of standard proportions are required, but size is variable, to be adjusted for yield strength and toughness of the materials.

3. E 561 Summary of Practice
   3.1 During slow-stable fracturing, the developing crack growth resistance, K_R is equal to crack extension force, K, applied to the specimen. The crack is driven forward by increments of increased load or displacement. Measurements are made at each increment for calculation of K values which are individual data points lying on the R-Curve of the material.
   3.2 The crack starter is a low stress level fatigue crack.
   3.3 Methods of measuring crack growth and of making plastic zone corrections to the physical crack length are prescribed. Expressions for the calculation of crack extension force are given.

4. E 561 Significance
   4.1 R-Curves characterize the resistance to fracture of materials during incremental slow-stable crack extension and result from growth of the plastic zone as the crack extends from a sharp crack. They provide a record of the toughness development as a crack is driven stably under increasing crack-extension forces.
   4.2 For an untested geometry, the R-Curve can be matched with the crack extension force curves to estimate the load necessary to cause unstable crack propagation. In making this estimate, R-Curves are regarded as though they are independent of starting crack size, a_0, and the specimen configuration in which they are developed. They appear to be a function of crack extension, Δa. To predict crack instability in a component, the R-Curve may be positioned as in Fig. 1, so that the origin coincides with the assumed initial crack length, a_0. Crack-extension force curves

for a given configuration can be generated by assuming applied loads or stresses and calculating crack extension force, K, as a function of crack size using the appropriate expression for K of the configuration. The unique curve that develops tangency with the R-Curve defines the critical load or stress that will cause onset of unstable fracturing.

4.3 If the K-gradient (slope of the crack extension force curve) of the specimen chosen to develop an R-Curve has negative crack drive slope, Fig. 2, as in the crack line wedge loaded specimen of method E 561, it may be possible to drive the crack until a maximum or plateau toughness level is reached. When a specimen with positive K-gradient characteristics is used, the extent of the R-Curve which can be developed is terminated when the crack becomes unstable.

5. E 561 Terminology

5.1.1.2 "Original crack size, a_0"(L)—the physical crack size at the start of the test.

5.1.1.3 "Effective crack size, a_e,"(L)—the physical crack size augmented for the effects of crack-tip plastic deformation.

5.1.5 "Crack extension resistance, $K_R(FL^{-3/2})$, and G_R or $J_R(FL-3/2)$"—a measure of the resistance to crack extension expressed in terms of the stress-intensity factor, K, the crack extension force, G, or values of J derived using the J-integral concept.

8. E 561 Procedure

8.3 "Loading Procedure"—load the CCT, CT, and CLWL specimens incrementally, allowing time between steps for the crack to stabilize before measuring load and crack length. Cracks stabilize in most materials within seconds of stopping the loading. However, when stopping at or near an instability condition, the crack may take several minutes to stabilize, depending on the stiffness of the loading frame and other factors.

8.5 "Effective Crack-Length Measurement"—compliance measurements, $2v/P$, made during the loading of specimens, can be used to determine effective crack length, a_e, directly. The crack is automatically plastic zone corrected and these values can be used directly in the expression for K.

9. E 561 Calculation and Interpretation

9.1 To develop a K_R-Curve, generate and use crack length and load data to calculate K_R.

9.1.1 For the center cracked tension specimen use one of the two following and equally appropriate expressions:

$$K_R = \frac{P}{WB} \sqrt{a_e} \left[1.77 = 0.177 \left(\frac{2a_e}{W}\right) + 1.77 \left(\frac{2a_e}{W}\right)^2 \right]$$

$$K_R = \frac{P}{WB} \left[\tau a_e \sec \left(\frac{\pi a_e}{W}\right) \right]^{1/2}$$

9.1.2 For the CT and CLWL specimens determine K_R as follows:

$$K_R = \frac{P}{B\sqrt{W}} \cdot f\left(\frac{a_e}{W}\right)$$

$$f\left(\frac{a_e}{W}\right) = \left[\frac{(2 + a_e/W)}{(1 - a_e/W)^{3/2}}\right] \left[0.886 + 4.64 \left(\frac{a_e}{W}\right) \right.$$
$$\left. - 13.32 \left(\frac{a_e}{W}\right)^2 + 14.72 \left(\frac{a_e}{W}\right)^3 - 5.6 \left(\frac{a_e}{W}\right)^4 \right]$$

9.1.4 The crack length used in the expressions of 9.1.1 and 9.1.2 is the effective crack length, which is the total physical crack length, a_p, plus a correction for plastic zone, r_Y. Correct physically measured crack lengths as follows:

$$a_e = (a_0 + \Delta a_p + r_Y)$$

10. E 561 Compliance Methods

10.1 The compliance technique uses elastic spring characteristics of the specimen calibrated over various crack sizes. A calibration curve may be developed experimentally by elastically loading specimens of varied crack sizes and determining the elastical reciprocal spring constant or reciprocal slope of the load versus displacement record. Normalize these slopes for material thickness and elastic modulus and plot against crack length to specimen width ratio. Analytically developed expressions for the compliance of common specimen geometries are available.

10.3 In testing to develop an R-Curve, the test record of load versus clip-gage displacement for the CCT and CT test or the $2v1$ versus $2v2$ test record will have an initial linear portion, the slope of which should correspond to the starting crack length in the specimen.

10.6 Calculate K_R in accordance with expressions given in paragraphs 9.1.1 and 9.1.2 using compliance determined effective crack length.

11. E 561 Report
      11.1.1 Type and size of specimen used
      11.1.2 Crack propagation direction in the material
      11.1.3 Material thickness
      11.1.4 Yield Strength
   11.2 The R-Curve may be plotted in terms of either physical or effective crack extension. The legend shall contain the following information; (*a*) Method of plastic zone adjustment; (*b*) whether the abscissa is Δa (effective) or Δa (physical). Instability predictions can be made only from effective crack extension plots.

Prediction Methodology

The K_R-curve is plotted on a K versus crack size coordinate system, placing the origin at the assumed initial crack size in the component to be predicted (Fig. 1). Again, crack growth is plotted in terms of effective crack extension.

The linear-elastic K_I solution for the geometry to be predicted is obtained. If load control is assumed, the family of crack drive curves is developed using various fixed load levels, calculating K_I as a function of crack size dimension for each level. The unique K_I versus crack size plot that develops tangency with the K_R-curve indicates instability or maximum load. It is required that the stress in the remaining ligament be determined at each assumed load level for supplementary information. This is compared with limit load (plastic collapse stress) based upon the crack size at the intersection point with the K_R-curve. Limit load is calculated using physical crack size inclusive of stable growth, and it is necessary to use the original K_R-curve data to determine the corresponding Δa_p at

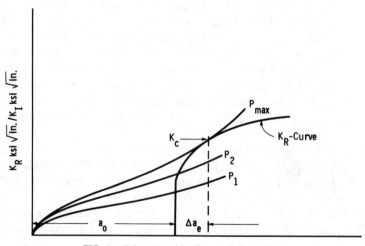

FIG. 1—*Schematic of load-controlled instability.*

the K_R level of intersection. Predictions by K_R-curve practice cannot be made for conditions that exceed onset of limit load deformation.

For displacement control assumption, crack instability may occur beyond maximum load. For an analysis, it is necessary to know the overall elastic compliance characteristics of the component between the load application points: $EB(2v/P) = f(a/W)$, where $f(a/W)$ represents the functional relationship between crack size and elastic spring behavior as a composite system. The terms in the compliance equation are reordered and substituted in the stress-intensity equation such that the load variable is replaced and K_I crack drive is expressed as a function of applied displacement instead of applied load. In a similar manner as before, various levels of applied displacement are assumed and K_I is calculated as a function of crack size. Figure 2 is a rather extreme illustration for a very rigid specimen. Most test systems are intermediate between the pure load control of Fig. 1 and rigid displacement condition of Fig. 2. Again, if a unique K_I versus a crack drive plot is shown that develops tangency with the K_R-curve, this predicts an instability condition. This may be at maximum load as before or it may be at some point beyond maximum load, under decreasing load conditions.

To put the present methodology into perspective, the approach taken is not different than the J_R-Δa_p R-curve prediction methodology [5] nor the tearing modulus (T-parameter) methodology, [6]. The predicted results will be the same, independent of choice so long as there is a demonstrated equivalency between deformation theory J_R and plasticity corrected K_R. This equivalency gradually degenerates after a limit load condition is surpassed. Hence, for highly ductile materials, the methodology is good only to predict maximum load. Again, care must be taken to compare limit load stress for the various assumed crack drive load levels.

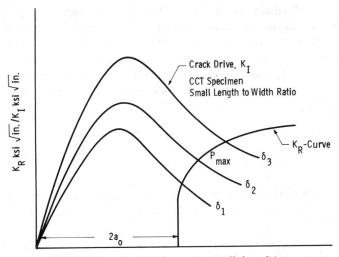

FIG. 2—*Schematic of displacement-controlled conditions.*

Limitations

The unique characteristics of specimen/component behavior that result in the equivalency between J_R and plasticity corrected K_R require the necessary and sufficient condition than η elastic and η for elastic-plastic deformation be reasonably equal. See Ref 7 for discussion of η. This has been shown to be essentially obeyed for most common material-geometry conditions [7] that display power hardening behavior. Nevertheless, a proper precaution would be to demonstrate that this equivalent η behavior is satisfied for the geometry being predicted.

R-curve development according to Method E 561 is a static crack method, requiring K_R and a_e determinations under stable-static crack conditions with a certain reasonable waiting period between each loading increment. Ductile materials when loaded to large plastic deformation conditions tend to show some time-dependent load relaxation characteristics. This introduces the possibility of slightly underpredicting the K_R-curve level of a continuously loaded structure, and hence a conservative maximum load prediction may result.

For high-strength material, the R-curve depends on the constraint of the specimen or component. If the constraint changes with geometry (thickness, a/W, plan view size, tension-bending loading), the R-curve will change according to the effect of these variables. On the other hand, if plane-stress conditions prevail, then there is experimental evidence that the R-curve is independent of geometry variables like a/W, plan view size, and specimen loading mode (bending or tension configurations) [8]. In other words, we can predict the behavior of a center-cracked tension (CCT) specimen from a compact tension (CT) or single-edge notched bending (SENB) specimen.

The highly ductile grades of structural materials have not performed quite as predictably. Experimental evidence exists that J_R-curve basic geometry independence like CCT versus CT can be lost. Within a basic geometry, the J_R-curve or K_R-curve will continue to be independent of initial crack size and specimen plan view size, again provided conditions of constraint are not varied within these specific dimensional differences. The Δa_e characteristic is specimen size independent as illustrated in Fig. 3 for a wide range of compact specimen sizes. Although the K_R-J_R equivalency is maintained within the given geometry, the K_R-curves are not shown to be transferable over different loading modes (bend versus tension).

Example Calculations

Problem No. 1

The first prediction represents a clear-cut example matchup between linear elastic crack drive and K_R-curve. Maximum stress is predicted for a three-hole panel of 7075-T651 aluminum. The panel is 25 cm wide, 12.7 mm thick, and contains a crack of 65 mm as defined in Fig. 4. The pertinent K_R-curve was given in a E24.06.02 round-robin activity and was obtained from a 4T compact

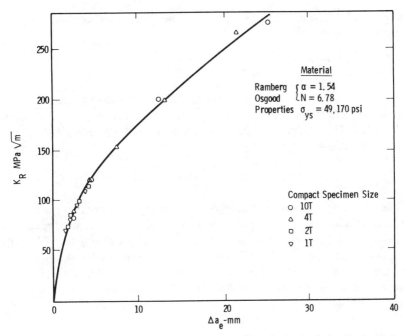

FIG. 3—*Effective crack growth and K_R equivalent to J_R from elastic-plastic handbook solution.*

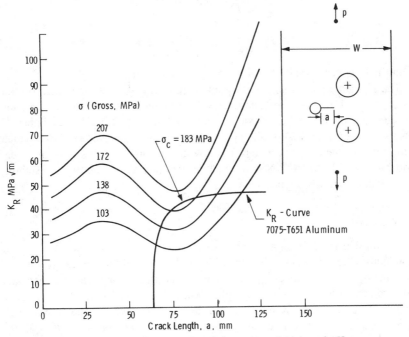

FIG. 4—*Instability prediction on three-hole specimen. Initial crack: 65 mm.*

specimen of 12.7-mm material thickness. The crack drive equation was also given:

$$K = \sigma \sqrt{\pi a}\, F \qquad (1)$$

Values for F were tabulated by Newman. Crack drive curves are shown for several levels of stress. The best estimated stress for tangency with the K_R-curve is 183 MPa. This compares to the experimental maximum stress of 192 MPa for about 5% prediction error.

Problem No. 2

The second problem is the prediction of a maximum or instability load (load control) for a tougher aluminum alloy in a 25-cm-wide center-cracked panel with $2a_0/W = 0.4$. This will be a case where limit load conditions interdict in the instability load prediction. The K_R-curve had also been developed in the previously mentioned ASTM E24.06.02 round-robin activity using a 4T plan view compact specimen with 12.5-mm-thick 2024-T351 aluminum. For the problem, assume the following: $W = 25.4$ cm, $B = 12.5$ mm, $a_0 = 50.8$ mm, $\sigma_{ys} = 314$ MPa, and $\sigma_u = 458$ MPa. See also Table 1.

$$K_I = \frac{P}{BW}\left(\pi a \sec \frac{\pi a}{W}\right)^{1/2} \qquad (2)$$

$$P_L = 1.1\, \sigma_{ys}\,(W - 2a_p)B \qquad (3)$$

Limit load is based on 110% of material yield strength.

The K_R-curve (set at $a_0 = 50.8$ mm) is plotted in Fig. 5, and crack driving force is shown for the five assumed levels of applied load. The K_R at the intersection is used to predict physical crack growth, Δa_p, from the available 4T compact data (of the E24.06.02 report). The resulting a_p is then used to calculate limit load for the CCT specimen geometry (last column). In this case, limit load capability of 633 to 643 kN is reached before crack drive controlled instability can occur at 672-kN applied load.

Problem No. 3

The third example is for a center-cracked panel of 2024-T351 aluminum, again nominally 12.5 mm thick, 40.6 cm wide, with an initial total crack size of 15.7 cm. In this case, however, the panel is loaded under displacement-controlled conditions and the test system compliance must be considered. The dimensions and displacement measurement locations are shown in Fig. 6. The problem here is to calculate the critical applied displacement for instability.

TABLE 1

Applied Load, kN	Crack Drive for 25-cm-Wide CCT Specimen K_I MPa $\sqrt{m}$ at a = (mm)								K_R-Curve Intersection		Limit Load P_L (kN)
	(25.4)	(50.8)	(57)	(63.5)	(70)	(76.2)	(82.5)	(88.9)	MPa $\sqrt{m}$	a_p (mm)	
334	30	46	50	55	60	...	...	...	47	51	668
556	50	77	84	92	100	...	...	...	86	53	649
623	56	86	94	103	112	123	...	...	100	53.6	643
645	58	89	97	106	116	128	...	...	106	54.9	633
672	60	93	101	111	121	132	147	163	125	56.9	615

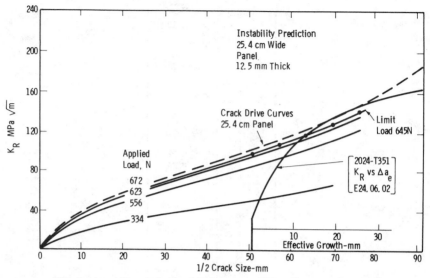

FIG. 5—*Instability prediction of 25.4-cm-wide CCT specimen of 2024-T351.*

For crack drive we again employ

$$K_I = \frac{P}{BW} \left(\pi a \sec \frac{\pi a}{W} \right)^{1/2} \tag{4}$$

Specimen compliance is given by

$$\frac{EB[2v]}{P} = 2\{(\pi a/W)/\sin(\pi a/W)\}^{1/2} \times F(Y/W, a/W) \tag{5a}$$

$$F = \frac{Y}{W} \left\{ (2W/\pi Y) \cosh^{-1}\left[\cosh(\pi Y/W)/\cos(\pi a/W)\right] \right.$$

$$\left. - \frac{1 + \mu}{\left[1 + \left(\dfrac{\sin \pi a/W}{\sinh \pi Y/W}\right)^2\right]^{1/2}} + \mu \right\} \tag{5b}$$

Displacement, $2v$, over a given panel length, $2Y$, is estimated from the above compliance equation.

Estimating the effective panel length (Fig. 6) to be 71 cm, $Y/W = 35.5/40.6$, and replacing P in Eq 4 with $P = EB[2v]/f(a/W, Y/W)$ from Eq 5, we obtain calibration $K/[2v] = f(a/W, Y/W)$. See Table 2.

Maximum load is predicted to be about 828 kN at $2v = 2.286$ mm. However,

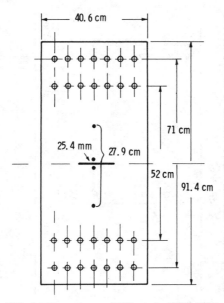

FIG. 6—*Dimensions on CCT panel of 2024-T351.*

instability at K_R = 123.2 MPa$\sqrt{\mathrm{m}}$ is predicted to be beyond maximum load at 770 kN (Fig. 7). The actual panel test had a maximum load of 801 kN and instability at 774 kN when $2v$ was 2.362 mm. To verify this prediction, the experimental displacement as instability was calculated as follows:

$$\delta_{\mathrm{total}} = \text{(measured displacement over 279.4-mm gage length)} + \sigma(L - 279.4)/E \text{ (see Fig. 6)}$$

$$= 1.7653 + \frac{P}{BW} \frac{(L - 279.4)}{E}$$

$$\delta_T = 1.7653 + \frac{774 \times 10^3 \times (L - 279.4)}{12.5 \times 406.4 \times 68.94 \times 10^3}$$

For L = 710 mm

$$\delta_T = 1.7653 + 0.954 = 2.719 \text{ mm}$$

Comparing this to the calculated value of 2.36 mm suggests that the effective elastic length of the panel was short of the last row of bolts. Recalculating at L = 520 mm for the first row of bolts gives δ_T = 2.3 mm, which means that the experimental effective elastic length of the panel lies between the first and second row of bolts.

TABLE 2

a/W	$EB2v/P$	$K/[2v] \times 10^{-6}$ MPa$\sqrt{m}$	K_I, MPa $\sqrt{m}$		
			$2v = 2.03$ mm	$2v = 2.286$ mm	$2v = 2.362$ mm
0.15	1.937	40.6	82.6	92.9	96
0.19	2.060	44.7	90.75	...	...
0.20	2.097	45.57	92.51	104	107.4
0.22	2.178	47.14	95.7	...	...
0.24	2.271	48.77	99	111	114.5
0.26	2.377	49.8	101.1	...	...
0.28	2.498	50.97	103.5	116.4	120.3
0.30	2.636	52.05		...	...
0.32	2.795	53.09	...	124.3	...
			at $2v = 2.03$ mm	at $2v = 2.286$ mm	at $2v = 2.362$ mm
K_R (MPa $\sqrt{m}$)			97.9	114.4	123.2
a_c (cm)			9.75	10.82	12.44
P (kN)			799	828	770

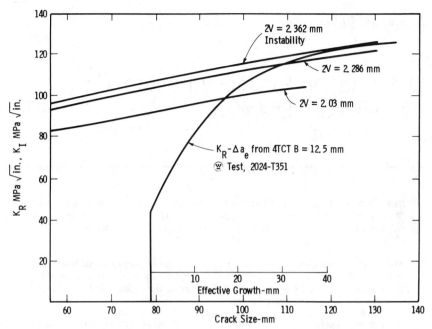

FIG. 7—*Displacement-control-loaded 40.6-cm-wide CCT panel of 2024-T351.*

References

[1] McCabe, D. E. and Landes, J. D., "An Evaluation of Elastic-Plastic Methods Applied to Crack Growth Resistance Measurements," *Elastic-Plastic Fracture, ASTM STP 668*, American Society for Testing and Materials, Philadelphia, 1979, pp. 286–306.

[2] Schwalbe, K. H. and Setz, W., "R-Curve and Fracture Toughness of Thin Sheet Materials," *Journal of Testing and Evaluation,* July 1981, pp. 182–194.

[3] McCabe, D. E. and Landes, J. D., "Elastic-Plastic R-Curves," *Journal of Engineering Materials and Technology, Transactions,* American Society of Mechanical Engineers, Vol. 100, April 1978.

[4] McCabe, D. E. and Ernst, H. A., "A Perspective on R-Curves and Instability Theory," *Fracture Mechanics: Fourteenth Symposium, Volume I: Theory and Analysis, ASTM STP 791*, Vol. 1, American Society for Testing and Materials, Philadelphia, 1981, pp. I-561–I-584.

[5] Shih, C. F. and German, M. D., "An Engineering Approach for Examining Crack Growth and Stability in Flawed Structures," G. E. Report 80CRD 205, General Electric Co., Schenectady, NY, Sept. 1980.

[6] Paris, P. C., Tada, H., Zahoor, Z., and Ernst, H. A., "Instability of the Tearing Mode of Elastic-Plastic Crack Growth," *Elastic-Plastic Fracture, ASTM STP 668*, American Society for Testing and Materials, Philadelphia, 1979, pp. 5–36.

[7] Turner, C. E., "The Ubiquitous Eta Factor," *Fracture Mechanics, ASTM STP 700*, American Society for Testing and Materials, Philadelphia, 1980, pp. 314–337.

[8] Hellmann, D. and Schwalbe, K. H., "Geometry and Size Effects on J_R and δ_R Curves Under Plane Stress Conditions, *Fifteenth Symposium, Fracture Mechanics:* American Society for Testing and Materials, Philadelphia, 1982, pp. 577–605.

Joseph M. Bloom[1]

Deformation Plasticity Failure Assessment Diagram

REFERENCE: Bloom, J. M., "**Deformation Plasticity Failure Assessment Diagram,**" *Elastic-Plastic Fracture Mechanics Technology, ASTM STP 896*, J. C. Newman, Jr., and F. J. Loss, Eds., American Society for Testing and Materials, Philadelphia, 1985, pp. 114–127.

ABSTRACT: An engineering procedure for assessing the integrity of flawed structures is presented. The procedure uses results given in a plastic fracture handbook developed by General Electric in the format of the Central Electricity Generating Board of the United Kingdom R-6 failure assessment diagram.

The failure assessment diagram recognizes both brittle fracture and net section collapse of the flawed structure. It is a safety/failure plane defined by the stress-intensity factor/fracture toughness ratio (K_r) as the ordinate and the applied stress/net section plastic collapse stress ratio (S_r) as the abscissa. For a particular stress level and defect size, the coordinates (S_r, K_r) can be easily calculated. If the assessment point lies inside the failure assessment curve, the structure is safe. The distance of the assessment point from the curve is a direct measure of the margin of safety of the assessed structure.

The procedure can handle ductile tearing by redefining the failure assessment curve as the boundary between stable and unstable crack growth. Application of this approach to 2024-T351 aluminum center-cracked panels using plane-stress J_R resistance curves developed from compact specimens has given predictions of maximum load to within 2% of experimental values.

An additional example problem of a cracked pressure vessel is included which demonstrates the applicability of this procedure to real structures. Lastly, limitations of the deformation plasticity failure assessment diagram are discussed.

KEY WORDS: deformation plasticity failure assessment diagram, fracture mechanics, tearing instability, stable crack growth, J-R curves, plasticity, elastic-plastic fracture

A simple engineering procedure, referred to as the deformation plasticity failure assessment diagram (DPFAD) approach, for the prediction of instability loads is presented. The DPFAD recognizes both brittle fracture and net section collapse of a flawed structure or test specimen. It is a safety/failure plane defined by the stress-intensity factor/fracture toughness ratio as the ordinate and the applied stress/net section plastic collapse stress ratio as the abscissa. For a particular

[1] Technical advisor, Structural Mechanics, Babcock & Wilcox, Research and Development Division, Alliance, OH 44601.

stress level and defect size, these coordinates can be readily calculated. If the assessment point lies inside the failure assessment curve, no crack growth can occur. If the assessment point lies on the assessment curve, stable crack growth is possible. If the assessment point lies outside the curve, unstable crack growth is predicted.

The basis of this approach began with work sponsored by the Electric Power Research Institute (EPRI) in the development and selection of a basic theory that describes the ductile fracture process. This initial work was done primarily by the General Electric Co. [1] and Battelle Columbus Laboratories [2]. These studies concluded that the J-integral is a valid parameter for characterizing crack initiation and growth, and that stable ductile tearing (slow crack growth) and load instability can be treated by a resistance curve approach [3]. Additional EPRI-funded work developed fully plastic solutions (deformation plasticity) for fracture mechanics test specimens, flawed cylindrical configurations, and a simple nozzle geometry [4]. The DPFAD, in the format of the Central Electricity Generating Board's (CEGB) R-6 failure assessment diagram [5], uses the results of deformation plasticity solutions developed by General Electric [4]. The CEGB R-6 failure assessment approach was derived from the original CEGB two-criteria approach [6]. Both CEGB approaches state that structures can fail by either of two mechanisms, brittle fracture or plastic collapse, and that these two mechanisms are connected by a interpolation curve based on the strip yield model [7]. This enables the analyst to go directly from linear elastic fracture mechanics (LEFM) calculations to plastic instability calculations. The DPFAD, however, is more accurate since it accounts for the actual material tensile properties, as well as the geometry of the flawed structure. This is because the DPFAD is based on deformation plasticity, the J-integral estimation scheme, and solutions from the *Plastic Handbook* [4]. This inherent accuracy of the DPFAD allows prediction of both fracture initiation and load instability when used in conjunction with the J-integral-based resistance curve approach [1,2].

General DPFAD Approach

The DPFAD approach for predicting instability loads for both test specimens and structures consists of the following steps.

DPFAD Curve Generation

The general approach is to first obtain the J-integral response for the flawed specimen/structure of interest. If a material can be modeled by deformation plasticity and its stress-strain behavior represented by a power-law strain-hardening equation, then simple expressions for the J-integral structural response can be written in terms of the power-law strain-hardening exponent (*n*). For a power-law material, Shih and Hutchinson [8] first showed that an estimate of a flawed structure's behavior for the complete range of applied stress can be formulated approximately by

$$J = J_1^e \, (a_{\text{eff}}, P) + J^P \, (a, P, n) \tag{1}$$

where

J_1^e = elastic contribution based on Irwin's plasticity-adjusted crack size (a_{eff}),
a = physical crack size,
P = applied remote load, and
J^P = deformation plasticity or fully plastic solution.

J_1^e expressions are available from various elastic fracture handbooks and J^P solutions can be found in Ref 4.

The DPFAD curve expression is obtained by normalizing the sum of the elastic and plastic response by the "elastic" J-integral of the structure in terms of a, where

$$J_1^e(a) = (1 - v^2)/E \, K_I^2(a) \tag{2}$$

and K_I is the LEFM stress-intensity factor. E and v are Young's modulus and Poisson's ratio, respectively. The normalized J-response is then defined by

$$K_r = \sqrt{J_1^e/J} = f(S_r) \tag{3}$$

where

$$S_r = \sigma/\sigma_L(a) \tag{4}$$

σ is the remote applied stress and σ_L is the reference plastic collapse stress or limit stress, a function of a and the material yield strength, σ_0.

Equation 3 defines a curve which is a function of the flaw geometry, structural configuration, and the stress-strain behavior of the material of interest.

Assessment Point Evaluation

To determine the instability load of a flawed structure, a locus of assessment points corresponding to some postulated stable crack growth must first be calculated in terms of K_r, S_r coordinates. The assessment point coordinates will be denoted by K'_r, S'_r to differentiate them from the K_r, S_r coordinates of the DPFAD curve defined by Eq 3. For stable crack growth, K'_r and S'_r are defined by

$$K'_r \, (a_0 + \Delta a) = \sqrt{J_1^e \, (a_0 + \Delta a)/J_R \, (\Delta a)} \tag{5}$$

and

$$S'_r \, (a_0 + \Delta a) = \sigma/\sigma_L \, (a_0 + \Delta a) \tag{6}$$

where J_I^e, J_R, and σ_L are all functions of the amount of postulated stable crack growth. The reference plastic collapse stress, σ_L, is now a function of the current crack size, $a_0 + \Delta a$. For actual crack initiation, $J \geqslant J_{Ic}$ [where J_{Ic} is determined from the ASTM Test Method for J_{Ic}, a Measure of Fracture Toughness (E 813-81)], Eq 3 defines the boundary between no crack growth and crack growth. For actual stable crack growth, $J = J_R$ (where J_R is obtained from the experimentally measured J_I-R data [9])

$$K'_r \cong K_r$$

and   (7)

$$S'_r \cong S_r$$

and the curve defined by Eq 3 becomes the stable crack growth path in the K_r, S_r plane, as shown schematically in Fig. 1. If $J > J_R$, load instability results and the crack growth path goes outside the K_r-S_r curve for load-controlled structures.

For an assumed remote constant tensile load of 445 kN (100 kips) and material properties as given in Table 1, the locus of assessment points for postulated crack extensions of 12.7 mm (50 mils) denoted by 1 to 127.0 mm (500 mils) denoted by 6 is shown in Fig. 2 for a 254-mm-wide (10-in.) center-cracked tension specimen of 2024-T351 aluminum 12.6 mm (0.495 in.) thick with an initial crack, $2a_0$, of length 102 mm (4 in.). The material's J_I-R curve is shown in Fig. 3.

In most applications where $\Delta a << a_0$, the DPFAD curve is defined by Eq 3 for $a = a_0$ (the initial crack size). This DPFAD curve is a conservative lower-bound approximation to the exact failure assessment diagram curve where $a = a_0 + \Delta a$.

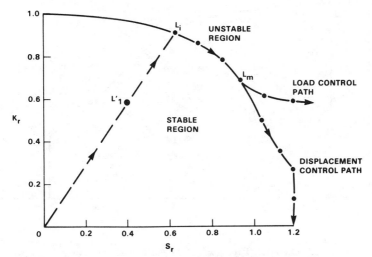

FIG. 1—*Failure assessment diagram for stable crack growth.*

TABLE 1—*Numerical results for example problem of 254-mm-wide (10 in.) center-cracked tension specimen (2 $a_0/w = 0.4$) of 2024-T351 aluminum; thickness = 12.6 mm (0.495 in). $\bar{P} = 100$ kips*

Point	Δa, in.	J_R, in. · lb/in.^{2}	S'_r	K'_r	Load, kips
1	0.050	500	0.856	0.797	111.6
2	0.100	750	0.871	0.663	123.9
3	0.200	1025	0.902	0.587	129.9
4	0.300	1300	0.935	0.541	130.6
5	0.400	1525	0.971	0.517	129.1
6	0.500	1650	1.010	0.515	125.9

25.4 mm = 1 in.; 445 kN = 100 kips; 1 in.-lb/in.^{2} = 175.1 N-m/m^2.

Instability Load Prediction

Instability load or maximum load prediction of a flawed test specimen or structure can be determined by ratioing the distance from the origin of the diagram to the DPFAD curve passing through the assessment point farthest from the curve by the distance to the assessment point itself. This ratio of distances times the assumed constant load (used in calculating the locus of assessment points) gives the equilibrium load (where crack growth is stable) of the flawed specimen or structure. Table 1 presents the results of such a calculation for a center-cracked panel. The locus of assessment points, the labeled points in Fig. 2, represents postulated crack growth at a constant load level of 445 kN (100 kips). The point farthest from the assessment curve is Point 4. The ratio of OB/OA (1.306) for

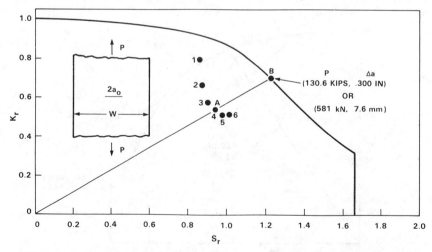

FIG. 2—*Failure assessment of a 2024-T351 aluminum center-cracked tension specimen. $\alpha = 0.25$, $n = 8.5$, $\sigma_{YIELD} = 276$ kN/m^2 (40 ksi), $a_0/w = 0.20$. Plane stress, P = 445 kN (100 kips), h_1 (0.4, 8.5) ≃ 1.80.*

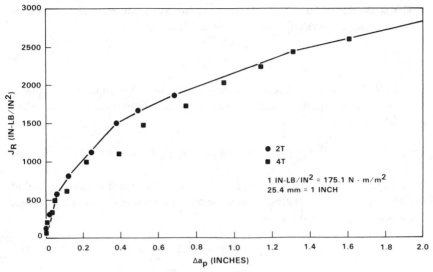

FIG. 3—*J_R-curve for 2024-T351 aluminum*

this point is multiplied by 445 kN (100 kips) to give a predicted instability load of 581 kN (130.6 kips) for the center-cracked panel. From Table 1 it can be seen that this instability load was predicted to occur at approximately 76.2 mm (300 mils) of crack growth. The experimental failure load for this specimen was 574 kN (129 kips).

Material Properties

The generation of the DPFAD curves requires a knowledge of the uniaxial *true* stress-*true* strain properties of the structure's material at the test temperature of interest.

The calculation of the locus of assessment points requires J_I-R curve data taken from compact specimen testing at the test temperature of interest per an established test procedure; that is, Ref 9 for plane-strain testing conditions. For plane-stress testing conditions, it is suggested that the compact specimen be of the same thickness as the structure.

Stress-Strain Tensile Properties

The required mechanical properties for the generation of the DPFAD curves are: Young's modulus (E), Poisson's ratio (ν), yield stress (σ_0), ultimate stress (σ_u), and the Ramberg-Osgood strain-hardening constants α, n defined by the Ramberg-Osgood stress-strain equation

$$\epsilon/\epsilon_0 = \sigma/\sigma_0 + \alpha(\sigma/\sigma_0)^n \tag{8}$$

The stress and strain (σ_0, ϵ_0) are chosen for convenience as the engineering yield strength, and the reference yield strain is defined by

$$\epsilon_0 = \sigma_0/E \qquad (9)$$

The constants α, n are determined by a least-squares best fit of the *true* plastic stress-*true* plastic strain plotted on log-log paper. The constants (α, n) which give the best fit to

$$\log (\epsilon_p/\epsilon_0) = \log \alpha + n \log (\sigma/\sigma_0) \qquad (10)$$

are used in the derivation of the expressions for the deformation plasticity failure assessment curves. The expression ϵ_p is the plastic portion of the Ramberg-Osgood stress-strain equation where

$$\epsilon_p = \epsilon - \sigma/E = \alpha \epsilon_0 (\sigma/\sigma_0)^n \qquad (11)$$

The *true* stress-*true* strain quantities are obtained from the engineering stress-strain quantities ($\bar{\sigma}$, $\bar{e}$) by

$$\epsilon = \log (\bar{e} + 1)$$
$$\sigma = \bar{\sigma}(\bar{e} + 1) \qquad (12)$$

J_I-R Curve

The material toughness data used in the calculation of the locus of assessment points (Eqs 5 and 6) can be obtained from a single compact specimen of the material at the desired test temperature. For plane-strain conditions, it is suggested that the test procedure presented in Ref 9 be used. For compact specimens 102 mm (4 in.) thick or greater, the J-expression found in ASTM Method E 813-81 can be used. For compact specimens where the amount of stable crack growth required in the analysis is less than 0.1 of the remaining ligament ($W-a$), deformation J is valid and the J-expression found in E 813-81 can be used. For compact specimens where the amount of stable crack growth required is greater than 0.1 of the remaining ligament (but less than 0.4 of the remaining ligament), it is suggested that the modified J-expression proposed by Ernst [10] be used. For specimen/structural thicknesses less than 25.4 mm (1 in.) where plane-stress conditions are possible, it is suggested that a "planar form" compact specimen configuration of the same thickness as the structure be used to determine the J-resistance toughness curve. A test procedure for this type of specimen is given in Ref 11. Test data in terms of J_R (Δa) obtained using the procedures of this reference were used in the analysis of the 2024-T351 aluminum 12.7-mm-thick (½ in.) center-cracked tension panels discussed in the next section.

Structural Applications

As part of the description of the DPFAD approach, two example problems are presented. The first example illustrates the prediction of the instability load of a 2024-T351 aluminum center-cracked tension specimen (AH6) using the fracture toughness results obtained from the testing of compact specimens reported in Ref *11*. The prediction is compared with the actual experimental result. Results of the prediction of three other similar specimens are also presented. The second example illustrates the determination of the factor of safety on pressure for a model of a pressurized water reactor (PWR) vessel under normal full-power operation. The vessel has a postulated continuous (full-length) longitudinal flaw in its beltline region. The initial flaw is assumed to be 25% of the vessel wall thickness. Material properties were taken from published data on 4T compact A533B steel specimens [*1*].

Center-Cracked Tension Specimen

The first step in the prediction of the instability load of the 254-mm-wide (10 in.) center-cracked tension specimen was to generate the DPFAD curve given the stress-strain properties of the 2024-T351 aluminum. The engineering stress-strain curve given in Ref *11* was converted to *true* stress-*true* strain and the plastic *true* stress was plotted versus the *true* strain on log-log paper. The resulting least-squares fit over the engineering strain range from 0.004 to 0.165 produced Ramberg-Osgood constants of $\alpha = 0.25$, $n = 8.5$, where σ_0 was chosen as 276 MPa (40 ksi).

The J-integral structural response expression reduced to the following DPFAD equation in terms of $K_r - S_r$:

$$J/J_1^e = 1/K_r^2 = a_e/a \cdot \frac{\sec(\pi a_e/a \cdot a/w)}{F_1^2} + \frac{\alpha h_1 S_r^{n-1}}{\pi F_1^2 (1 - 2a/w)} \quad (13)$$

where

$$a_e/a = 1 + \frac{1}{2}\left(\frac{n-1}{n+1}\right)\left[\sec\left(\frac{\pi a}{w}\right)\right]\frac{(1 - 2a/w)^2 S_r^2}{(1 + S_r^2)} \quad (14)$$

and

$$F_1^2 = \sec(\pi a/w)$$

The constant h_1 ($a_0/w = 0.2$, $n = 8.5$) $= 1.80$ was determined from Ref *4*. The vertical cutoff of the DPFAD curve shown in Fig. 2 was obtained from the ratio of the true ultimate strength divided by the reference yield strength, σ_0.

The second step of the instability prediction was to calculate a locus of assessment points per an assumed constant load of $P = 455$ kN (100 kips) for various postulated amounts of stable crack growth using the J_1-R toughness data

TABLE 2—*Prediction of P$_{max}$ for the ASTM E24.06.02 round-robin 2024-T351 aluminum center-cracked tension specimens.*

Specimen No.	B, in.	W, in.	a$_0$, in.	P$_{max}$, Maximum Failure Load, kips	
				Experimental	Predicted
AH91	0.495	5	1.030	67.9	68.8
AH92	0.495	5	0.992	70.0	71.4
AH3	0.495	10	2.016	130.7	133.7
AH6	0.495	10	2.050	129.0	130.6

1 kip = 4.448 kN; 1 in. = 25.4 mm.

shown plotted in Fig. 2. With these toughness data (connected by the straight lines), values of S'_r and K'_r were determined using Eqs 2, 5, and 6, where

$$K_1 = \sigma\sqrt{\pi a \, \sec{(\pi a/w)}} \qquad (15)$$

and

$$\sigma_L = (1 - 2a/w)\,\sigma_0 \qquad (16)$$

Table 1 presents the calculated assessment points per paired J_R-Δa toughness values. The points labeled 1 through 6 were then plotted as shown in Fig. 2. The last step determined the equilibrium load for each assessment point by ratioing the distance from the origin of the diagram to the DPFAD curve through each assessment point by the distance from the origin to the assessment point itself. This ratio was then multiplied by the assumed constant tension load to obtain the equilibrium load. For the maximum load or instability load, this ratio is given by OB/OA, as shown in Fig. 2. Note that the instability load of 581 kN (130.6 kips) compares quite well with the experimentally measured instability load of 574 kN (129.0 kips). Table 2 presents the results of similar predictions for three other center-cracked tension specimens. Note that the other predictions are equally good.

Axially Cracked Pressurized Cylinder

The object of this sample problem is to determine the factor of safety of a PWR vessel under normal full-power operation using the DPFAD approach. The vessel is assumed to have a continuous (full-length) longitudinal flaw of depth-to-wall thickness of $a/t = 0.25$.

The vessel is pressurized to 15.5 MPa (2.25 ksi) at a temperature of 250°C (480°F). The material properties were taken from EPRI/General Electric data on 4T compact A533B steel specimens [1].

Factor-of-safety calculations are made using the DPFAD for an axially cracked pressurized cylinder with a thickness-to-inside radius (t/R_i) of 0.10 and a continuous longitudinal flaw of a/t

The failure assessment curve expression for an axially pressurized cylinder is given by

$$1/K_r^2 = \frac{a_e}{a}\left[\frac{F(a_e)}{F(a)}\right]^2 + \frac{0.0026\,(1 - a/t)\,h'_1\,S_r^{\,n-1}}{\left[F(a)/10\,\dfrac{(1 - a/t)}{(1 + 0.1\,a/t)}\right]^2} \tag{17}$$

where

$$a_e/a = 1 + 22.22\left(\frac{n-1}{n+1}\right)\left[\frac{F(a)}{10}\,\frac{(1 - a/t)}{(1 - 0.1\,a/t)}\right]^2\,\frac{S_r^2}{(1 + S_r^2)} \tag{18}$$

and

$$\left[\frac{F(a_e)}{F(a)}\right]^2 = \left[\frac{1.165 - 1.339\,a_e/a\,a/t}{1.165 - 1.339\,a/t}\right]^2\left[\frac{1 - a/t}{1 - a_e/a\,a/t}\right]^5 \tag{19}$$

The vertical cutoff of the DPFAD curve shown in Fig. 4 was obtained from the ratio of the true ultimate strength divided by the reference yield strength, $\sigma_0 = 414$ MPa.

The material constants α, n, and the $h_1\,(a/t, n)$ calibration term are shown in Fig. 4. The assessment points are calculated using Eqs 5 and 6, where

$$J_1^e(a) = \frac{100\,p^2\,\pi a\,(1 - \nu^2)\,(1.165 - 1.339\,a/t)^2}{(1 - a/t)^5\,E} \tag{20}$$

and

$$p_1(a) = \frac{0.2}{\sqrt{3}}\,\sigma_0\,\frac{(1 - a/t)}{(1 + 0.1\,a/t)} \tag{21}$$

where $p_1(a)$ is the limit pressure based on yield strength, σ_0.

The point labeled No. 1 is the point corresponding to crack initiation (J_{Ic}), while Point No. 4 is the point corresponding to the maximum factor of safety. The factor of safety is determined by a line from the origin through the point of interest to the failure assessment curve. For initiation (see Fig. 4):

$$\text{Factor of safety} = OB/OA \tag{22}$$

The pressure required to initiate stable tearing is then 15.5 MPa (2.25 ksi) times OB/OA. The corresponding amounts of ductile tearing (Δa), equivalent J-resistance, S'_r, K'^2_r and factors of safety are given in Table 3. The factor of safety (rupture pressure divided by operating pressure) at initiation ($J = J_{Ic}$) is 1.71. The maximum factor of safety allowing for ductile tearing is 2.40.

[2] S'_r is now defined as the operating pressure divided by the limit pressure p_1 as given in Eq 21.

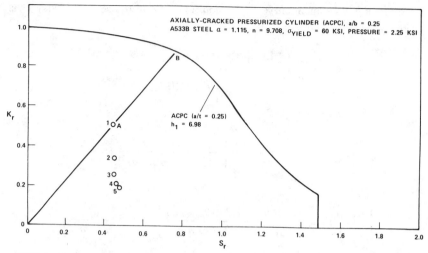

FIG. 4—*Failure assessment diagram of an axially cracked A533B steel pressurized cylinder.*

Limitations of Method

The DPFAD approach currently has three major limitations:

1. Accuracy of DPFAD curve (based on a fixed crack size) for the prediction of instability loads for large amounts of stable crack growth.
2. Limits of the validity of J_I-R curve data for deformation plasticity J-integral-based solutions.
3. Availability of fully plastic solutions for flawed structures of interest.

Fixed Crack Size for DPFAD Curve

The DPFAD procedure for the prediction of the instability load of a flawed structure is to plot the DPFAD curve for a fixed initial flaw size, a_0. However, for a stable growing crack the DPFAD curve changes, depending on the amount of stable crack growth to be considered. Figure 5 illustrates these inaccuracies

TABLE 3—*Numerical results for example problem of a pressurized cylinder with longitudinal crack (a/t = 0.25); p = 2.25 ksi (15.5 MPa).*

Point	Δa, in.	J_R, lb/in. $\times 10^3$	S'_r	K'_r	Factor of Safety
1	0.005	1.20	0.444	0.507	1.71
2	0.053	3.52	0.448	0.302	2.14
3	0.094	5.16	0.451	0.254	2.34
4	0.245	8.44	0.463	0.212	2.40
5	0.412	11.56	0.477	0.194	2.38

25.4 mm = 1 in.; MN/m = 5.71 $\times$ 10³ lb/in.
E = 27.3 $\times$ 10³ psi (188.2 $\times$ 10³ MPa); ν = 0.3.

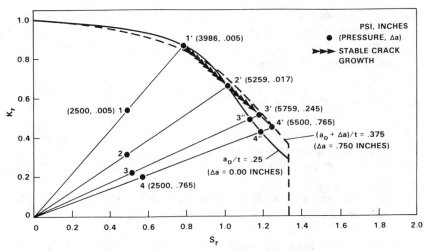

FIG. 5—*DPFAD curves for a_0 and $a_0 + \Delta a$ and indicated failure assessment of axially cracked pressurized cylinder. A533B steel:* $\alpha = 1$, $n = 10$, $\sigma_{YIELD} = 60\,ksi$, $\sigma_{ULTIMATE} = 80\,ksi$, $t/R_i = 0.10$, $a/t = 0.0$, $t = 8\,in.$, $a_0 = 2\,in.$

for an axially cracked pressurized cylinder which is postulated to have stable crack growth up to 750 mils (19 mm). Note that the actual equilibrium pressures for stable crack growth are determined by the ratios $01'/01$, $02'/02$, etc., times the postulated pressure of 17.2 MPa (2500 psi). However, from Fig. 5 it can be seen that for $\Delta a < a_0$ [19.05 mm < 50.8 (0.750 in. < 2 in.)], a conservative approximation is possible if the DPFAD curve is based on $a_0 = 50.8$ mm (2 in.) ($a_0/t = 0.25$), such that the calculated equilibrium pressures would be $03''/03 \times 17.2$ MPa (2500 psi), $04''/04 \times 17.2$ MPa (2500 psi), etc.

Limits of J_I-R Curve Data

All the predictive approaches based on the deformation plasticity J-integral are theoretically limited to the onset of crack extension. However, Hutchinson and Paris [12] showed that there is a J-controlled crack growth regime provided that certain crack growth criteria are met. These criteria are:

1. that the stable crack growth must be significantly less than the remaining specimen ligament, b; and
2. "ω" must be significantly greater than 1, where

$$\omega \equiv b/J \cdot dJ/da \tag{23}$$

where dJ/da is the slope of the J-resistance curve.

Recent work by Ernst [10] has shown that a modified version of the J-integral can be used to extend the validity of J_I-R curve data. While limited experimental data have demonstrated the validity of his work, further research is needed for his modification to gain full acceptance.

Availability of Fully Plastic Solutions

Current fully plastic solutions for J^p (see Eq 1) are limited to two-dimensional and axisymmetric flawed configurations, such as

- continuous (full-length) internal longitudinal flaws in cylinders under pressure loadings (plane strain conditions),
- continuous internal circumferential flaws in pressurized or axially loaded cylinders (axisymmetric conditions), or
- continuous external flaws in pressurized cylinders (either circumferential or axial; that is, plane-strain or axisymmetric conditions).

DPFAD curves using these models will be overly conservative because the stress-intensity factors for these flaws are much greater than those of a more realistic semi-elliptical flaw of the same depth. To date, only limited three-dimensional flawed structural models in terms of the DPFAD approach have been developed. In particular, Bloom [13] presented the solution of an American Society of Mechanical Engineers (ASME) Section III, Appendix G design flaw in a pressurized cylinder. This flaw is a semi-elliptical surface flaw with a maximum depth equal to one-fourth the wall thickness. The length of this flaw at the surface is 1.5 times the wall thickness.

Summary

An engineering approach for predicting instability loads for flawed test specimens and structures is presented. The procedure uses deformation plasticity solutions in the format of the Central Electricity Board of the United Kingdom R-6 failure assessment diagram, thereby giving it the name "deformation plasticity failure assessment diagram (DPFAD)." Determination of both the DPFAD curve and the failure assessment points are discussed and two example problems are presented. The first problem is that of an aluminum center-cracked panel, while the second is a pressurized flawed cylinder simulating a PWR vessel under normal operating pressure.

Lastly, current limitations of the DPFAD procedure are discussed.

References

[1] Shih, C. F. et al, "Methodology for Plastic Fracture," Final Report to Electric Power Research Institute, Contract No. RP601-2, NP-1735, General Electric Co., Schenectady, NY, Aug. 1980.

[2] Kanninen, M. F. et al, "Development of a Plastic Fracture Methodology," Final Report to Electric Power Research Institute, Contract No. RP601-1, NP-1734, Battelle Columbus Laboratories, Columbus, OH, March 1981.

[3] *Fracture Toughness Evaluation by R-Curve Methods, ASTM STP 527,* D. E. McCabe, Ed., American Society for Testing and Materials, Philadelphia, April 1973.

[4] Kumar, V. et al, "An Engineering Approach for Elastic-Plastic Fracture Analysis," EPRI Topical Report NP-1931, Research Project 1237-1, Electric Power Research Institute, Palo Alto, CA, July 1981.

[5] Harrison, R. P. et al, "Assessment of the Integrity of Structures Containing Defects," Report No. R/H/6, Central Electricity Generating Board, United Kingdom, 1976.

[6] Dowling, A. R. and Townley, C. H. A., "The Effects of Defects on Structural Failures: A Two-Criteria Approach," *International Journal of Pressure Vessels and Piping,* Vol. 3, 1975, p. 77.

[7] Dugdale, D. S., "Yielding of Steel Sheets Containing Slits," *Journal of the Mechanics and Physics of Solids,* Vol. 8, 1960, p. 100.

[8] Shih, C. F. and Hutchinson, J. W., "Fully Plastic Solutions and Large-Scale Yielding Estimate for Plane Stress Crack Problems," *Transactions, American Society of Mechanical Engineers, Journal of Engineering Materials and Technology,* Series H, Vol. 98, No. 4, Oct. 1976, p. 289.

[9] Albrecht, P., et al, "Tentative Test Procedure for Determining the Plane Strain J_I-R Curve," *Journal of Testing and Evaluation,* Vol. 10, No. 6, Nov. 1982, p. 245.

[10] Ernst, H. A., "Material Resistance and Instability Beyond J-Controlled Crack Growth," *Elastic-Plastic Fracture: Second Symposium,* Vol. 1, *ASTM STP 803,* C. F. Shih and G. P. Gudas, Eds., American Society for Testing and Materials, Philadelphia, 1983, p. I-191.

[11] "Predictive Round Robin on Fracture," ASTM Task Group E24.06.02 on Application of Fracture Analysis Methods, J. C. Newman, Jr., chairman, Aug. 1979.

[12] Hutchinson, J. W. and Paris, P. C. in *Elastic-Plastic Fracture, ASTM STP 668,* J. D. Landes, J. A. Begley, and G. A. Clarke, Eds., American Society for Testing and Materials, 1979, pp. 37–64.

[13] Bloom, J. M., "Extensions of the Failure Assessment Diagram Approach—Semi-Elliptical Flaw in Pressurized Cylinder," presented at the 1983 ASME Winter Annual Meeting, Boston, MA, preprint 83-WA/PVP-3, American Society of Mechanical Engineers, Nov. 1983; *ASME Journal of Pressure Vessel Technology,* Vol. 107, Feb. 1985, pp. 25–29.

Hugo A. Ernst[1] and John D. Landes[1]

Predictions of Instability Using the Modified J, J_M-Resistance Curve Approach

REFERENCE: Ernst, H. A. and Landes, J. D., **"Predictions of Instability Using the Modified J, J_M-Resistance Curve Approach,"** *Elastic-Plastic Fracture Mechanics Technology, ASTM STP 896,* J. C. Newman, Jr., and F. J. Loss, Eds., American Society for Testing and Materials, Philadelphia, 1985, pp. 128–138.

ABSTRACT: This paper presents a method for predicting the behavior of an untested structure using the modified J, J_M-resistance curve approach. The method combines the J_M-R curve, calibration functions, and the expression for J_M for the untested geometry to provide the load-load point displacement characteristics and the J_M-T_M diagram of the structure. These two records contain all the information needed to assess structural behavior, maximum load, the tendency for instability, and the load-point displacement at every point.

KEY WORDS: materials, structures, predictions instability, modified J, resistance curves

Nomenclature

a	Crack length
a_0	Initial crack length
$\Delta a = (a - a_0)$	Crack extension
b	Remaining ligament
b_0	Initial remaining ligament
B	Thickness
C	Linear elastic compliance of the cracked body
C_{CR}	Remaining compliance capacity
C_M	Compliance of structure
E'	Generalized Young's modulus
	$E' = E$ for plane strain and $E' = E/(1 + \nu)$ for plane stress
G	Linear elastic part of J or Griffith's energy release rate
J	Integral based on deformation theory of plasticity
J_{pl}	Plastic part of J
J_M	Modified J-integral

[1] Westinghouse R&D Center, Pittsburgh, PA 15235. Mr. Ernst is now with TECHINT, Buenos Aires, Argentina, and Mr. Landes with the American Welding Institute, Knoxville, TN.

J_{Mpl}	Plastic part of J_M
K	Stress-intensity factor
K_M	Linear elastic stiffness of a structure
n	Strain-hardening exponent $1 \leq n$
P	Load
T	Tearing modulus
T_{app}	Applied tearing modulus
T_{mat}	Material tearing modulus
T_M	Modified tearing modulus
T_{Mapp}	Applied modified tearing modulus
T_{Mmat}	Material modified tearing modulus
v	Load-point displacement
$v_{el} = PC$	Linear elastic part of displacement
$v_{pl} = (v-PC)$	Plastic part of displacement
$\dot{v}$	Displacement rate
W	Specimen width
σ_y	Yield stress
σ_u	Ultimate stress
$\sigma_0 = (\sigma_y + \sigma_u)/2$	Flow stress

Scope

The present methodology provides a way of predicting structural behavior of metallic cracked bodies based on results from laboratory tests. Basically, the method provides the entire load-load point displacement of the untested structure, that is, values of load, displacement and slope, dP/dv, at all points, and entire J_M versus T_M plots.

It is assumed here that a resistance curve can be obtained for the material and conditions of interest. Furthermore, it is assumed that the crack grows under ductile tearing; that is, cleavage fracture is excluded.

The method uses two pieces of information: the material resistance to crack growth, in terms of the J_M-Δa curve, and two calibration functions connecting the four variables of load, load-point displacement, crack length, and J in the geometry of interest.

Applicable Documents

ASTM Standards:

E 4	Load Verification of Testing Machines
E 8	Tension Testing of Metallic Materials
E 399	Test for Plane-Strain Fracture Toughness of Metallic Materials
E 616	Terminology Relating to Fracture Mechanics Methods
E 813	Test Method for J_{Ic}, a Measure of Fracture Toughness

Summary of the Method

This method serves to predict the behavior of an untested geometry, using the modified J, J_M [1,2] resistance curve approach, in terms of its load-load point displacement characteristics or J_M versus T_M diagram.

Two components are needed to apply this method: the material resistance to crack growth in terms of the J_M-Δa curve, and calibration functions of the geometry of interest, resulting from the elastic-plastic analysis, that is, two relations linking the four variables of load P, load point displacement v, crack length a, and J (ASTM E 813). For example

$$v = v(P,a)$$

$$J = J(P,a)$$

In addition, the expression for J_M for the untested geometry is needed, that is

$$J_M = J - \int_{a_0}^{a} \frac{\partial J_{pl}}{\partial a}\bigg|_{v_{pl}} da$$

where the term $\partial J_{pl}/\partial a$ can be calculated from the calibration functions.

The method then combines these components so as to find $(P, \Delta a)$ pairs for the new geometry that follow the J_M-Δa curve. These results are then converted to J_M-T_M diagrams or P-v records of the untested geometry. From these plots, all the relevant parameters can be obtained: the maximum load P_{max}, or instability load for load controlled conditions, the displacement at that point of instability v_{inst}, and the slope of the P-v record (dP/dv) at every point, which is related to the tendency that the structure has for instability [3].

From another point of view the modified applied tearing modulus T_{Mapp} can be calculated at every point and compared with the modified material tearing modulus T_{Mmat} to determine the point of instability using the J_M-T_M diagram. In conclusion, by combining the J_M-R curve, the calibration functions, and the expression for J_M for the untested geometry, this method provides all the needed information to assess structural performance under different loading conditions.

Significance

The load-load point displacement and J_M-T_M diagrams for an untested geometry obtained by this method provides all the information needed to assess structural reliability and material selection. The maximum load P_{max} represents the load-bearing capacity of the structure or the load at instability under load-controlled conditions. The load-point displacement at maximum load represents the displacement underwent by the load point at instability load under load-controlled conditions. The slope, $-dP/dv$, at any point, is related to the tendency for instability of the structure. Its inverse defines the remaining compliance capacity C_{CR} [3] which represents the minimum extra compliance needed to be added in series to cause instability at that point under total displacement-controlled conditions. That is

$$C_{CR} > - \left(\frac{dP}{dv}\right)^{-1}$$

for instability under displacement-controlled conditions.

Also, the mentioned fraction ($-dP/dv$) gives the minimum necessary stiffness K_M in parallel with the structure to prevent instability under load-controlled conditions. That is

$$K_M > - \frac{dP}{dv}$$

for stability under load-controlled conditions.

Furthermore, the energy applied to the structure needed to grow the crack to a certain level can be obtained by measuring the area under the $P\text{-}v$ record up to the point of interest.

Laboratory Tests

In general, the laboratory test procedure should follow the guidelines of Refs 4 and 5 except for the particular cases discussed below.

Test Specimen Design and Preparation

Follow Ref 4 and applicable ASTM Standards (E 4, E 8, E 399, E 616, E 813).

Test Procedure

Follow Ref 4 and applicable ASTM Standards except for the limits regarding allowable crack extension and ligament to J_M ratio. The values proposed here [1–6] are

$$\Delta a \leq 0.4 \, b_0$$

and

$$\frac{b \, \sigma_0}{J_M} \geq 5$$

Calculation and Interpretation of Results

Two quantities have to be determined from the laboratory tests, the crack extension Δa and J_M. The former is measured following Ref *4*; the latter is obtained using

$$J_M = J + \int_{a_0}^{a} \frac{m}{(W - a)} J_{pl} \, da$$

where $m = 1$ for the 3-point bend specimen (3PB) and $m = (1.76 - 0.76 \, a/W)$ for the compact specimen (CT). J is determined following Ref *5*.

Report Results

The objective of the laboratory test is to provide the material resistance to crack growth in terms of the J_M-R curve, that is, J_M-Δa pairs. The results thus should be reported in a table containing the following data for each step: J_M, Δa, P, v, v_{el}, J, and G, as well as information about the initial condition, that is, specimen dimensions a_0, and initial (measured) linear elastic compliance. Additional information should include temperature, deformation rate $\dot{v}$, and the tensile properties of the material such as the whole stress-strain curve.

Structural Applications

Analysis Required

To apply the present method, an elastic-plastic analysis is needed for the geometry of interest, capable of providing two equations connecting the four variables P, v, a, and J. For example, taking v and J as the dependent variables

$$v = v(a,P)$$

$$J = J(a,P)$$

or in more familiar form

$$v = PC + v_{pl}(a,P)$$
$$J = G + J_{pl}(a,P)$$

Procedure

The procedure to follow in applying this method is explained below in a step-by-step scheme.

1. Obtain tensile material properties, E, σ_y, n and the whole stress-strain curve.

2. Obtain material response to crack growth in terms of a J_M-R curve, that is, $(J_{Mi}, \Delta a_i = (a_i - a_0))$ pairs.

3. Specify the geometry of interest and the length parameters involved: W, B, others if any, as well as the initial crack length a_0.

4. Obtain the two calibration functions linking the four variables P, v, a, and J. For example

$$v = v(a,P)$$

$$J = J(a,P)$$

5. Consider the load $P = \Delta P$ and crack length $a_1 = a_0 + \Delta a_1$.

6. Calculate J for P and a_1

$$J(a,P) = J_{11}$$

7. Calculate the corresponding J_{M11} using

$$J_M = J - \int_{a_0}^{a} \frac{\partial J_{pl}}{\partial a}\bigg|_{v_{pl}} da$$

The term $\partial J_{pl}/\partial a$ is calculated from the calibration functions.

8. Compare J_{M11} with J_{M1}.

9. Keep on increasing P in steps ΔP until $P = P_1$ for which $J_{M11} = J_{M1}$.

10. Save that pair $(P_1, a_1 = a_0 + \Delta a_1)$.

11. Move to the next crack length $a_2 = a_0 + \Delta a_2$.

12. Increase the load in steps ΔP.

13. For each step calculate J as a function of load (P) and crack length a_2.

$$J_2 = J_2 (a_2,P)$$

14. Calculate J_{M2} for each step until a load P_2 is reached for which

$$J_{M2} = J_{M22}$$

15. Save the pair (P_2, a_2).

16. Continue up to the last Δa value.

In conclusion, pairs $(P_i \Delta a_i)$ have been found for which the associated values $(J, \Delta a)$ are such that when modified to $(J_M, \Delta a)$ agree with the J_M-R curve provided.

Calculation of Results

The structural performance can be assessed by two means, namely, the P-v record or the J_M-T_M diagram. Once the pairs $(P_i\ \Delta a_i)$ have been found for the structure, it is a straightforward to produce the two diagrams. The P-v record can be obtained just by using

$$v = v(P,a)$$

for each step.

On the other hand the J_M-T_M diagram, that is, J_M versus T_{Mmat} and J_M versus T_{Mapp}, can be obtained as follows. The former entails only the normalization of the J_M-Δa curve provided by the laboratory tests, that is

$$T_{Mmat} = \frac{E}{\sigma_0^2} \frac{dJ_M}{da}$$

An alternative formula can also be used if desired

$$T_{Mmat} = \frac{E}{\sigma_0^2} \left[\frac{\partial G}{\partial a} - \frac{\partial J}{\partial v}\bigg|_a \frac{\partial P}{\partial a}\bigg|_v \frac{1}{\left(\frac{\partial P}{\partial v}\bigg|_a - \frac{dP}{dv}\right)} \right]$$

T_{Mapp} can be obtained using

$$T_{Mapp} = \frac{E}{\sigma_0^2} \left[\frac{\partial G}{\partial a} - \frac{\partial J}{\partial v}\bigg|_a \frac{\partial P}{\partial a}\bigg|_v \frac{1}{\left(\frac{\partial P}{\partial v}\bigg|_a + K_M\right)} \right]$$

At the same time the negative value of the slope of the load displacement record $(-dP/dv)$ can be calculated at every point. Its inverse gives the minimum compliance needed to be added in series with the structure to cause instability under displacement-controlled conditions at that point [3]. This compliance has been defined as the remaining compliance capacity C_{CR}

$$C_{CR} > \left(-\frac{dP}{dv}\right)^{-1}$$

for instability under displacement-controlled conditions.

The value of $(-dP/dv)$ also gives, at every point, the minimum necessary stiffness K_M that a member in parallel with the structure has to have to prevent instability (even beyond maximum load).

Note that for the raising part of the *P-v* record, dP/dv is positive and then $-dP/dv$ is negative. That implies that there is no value of C_{CR} or K_M.

Report of Results

The results should be reported in terms of the *P-v* diagram and J_M versus T_M plots. A table containing the values of the variables of interest P, Δa, v, v_{el}, v_{pl}, G, J, and J_M at every step should be also included in the report.

Limitations of the Method

This method is based on three concepts:

1. That a resistance to crack growth curve can be obtained for the material and conditions of interest. The crack grows by stable tearing and cleavage fracture is excluded.

2. That calibration functions of the form $v = v(a,P)$ and $J = J(a,P)$ can be found for the geometry of interest.

3. That the crack growth mechanism and the constraint (that is, degree of plane strain versus plane stress) are the same in the test specimen and in the geometry of interest. If any of the above assumptions are not correct, the method is not expected to give accurate results.

An additional (obvious) point is that the prediction cannot be carried on further, in crack extension, than what is available from the resistance curve. Thus, it is important to have these curves as extended as possible in Δa.

Example Calculations

Problem I

Given the *J-Δa* curve and *P-v* record of a CT specimen of dimensions $W = 203.2$ mm (8 in.), $B = 6.35$ mm (0.25 in.), and $a_0 = 121.9$ mm (4.8 in.) of a 2024 aluminum alloy, calculate the maximum load P_{max} (or load at instability under load-controlled conditions) of a CCT specimen whose dimensions are total width $W = 254$ mm (10 in.), $B = 6.35$ mm (0.25 in.), and total crack length $a_0 = 10.41$ mm (4.10 in.). The material properties are $\sigma_y = 314$ MPa (45.5 ksi), $\sigma_u = 458$ MPa (66.4 ksi), and the hardening exponent $n = 6.69$.

Answer: The experimentally determined P_{max} was 574 kN (129 klb).

Solution

The *J-Δa* curve from the CT was converted to a J_M-Δa curve using the *P-v* record and the formula

$$J_M = J + \int_{a_0}^{a} \frac{(J - K^2/E)}{(w - a)} (1.76 - 0.76\, a/w) da$$

Once the J_M-Δa curve was obtained, the procedure of the previous sections was followed in detail, where J_M for the CCT is given by [1]

$$J_M = J - \int_{a_0}^{a} \frac{(J - K^2/E)}{(w - a)n} da$$

The *EPRI Handbook* [7] was used and plane-stress conditions were assumed, yielding a P_{max} of 571 kN (128.4 klb). Note the excellent agreement between the predicted value and the experimentally measured one. The error is 0.5%.

Problem II

Given the J-Δa curve, P-v record and material properties of a CT specimen whose dimensions are $W = 203.2$ mm (8 in.), $B = 181.6$ mm (4 in.), and $a_0/W = 0.55$. Calculate for a CT of $W = 508$ mm (20 in.), $B = 254$ mm (10 in.), and $a_0 = 254$ mm (10 in.) the following parameters.

1. Maximum load P_{max} or instability load under load-controlled conditions.
2. Displacement at P_{max}, v_{inst}.
3. If a spring is set in series with the specimen and the whole system is subjected to displacement-controlled conditions, what is the minimum spring compliance needed to cause unstable crack growth?
4. If a spring is set in parallel with the specimen and the whole system is subjected to load-controlled conditions, what is the minimum spring stiffness needed to prevent instability? The material properties are $\sigma_y = 366$ kN (53.14 ksi), $\sigma_u = 552$ kN (80 ksi), and the hardening exponent is $n = 7.13$.

Answers: The experimentally determined quantities are

$$P_{max} = 2936 \text{ kN (660 kips)}$$

$$v_{inst} = 6.6 \text{ mm (0.26 in.)}$$

Solution

The J-Δa curve from the CT specimen was converted to a J_M-Δa one using the P-v record and the formula

$$J_M = J + \int_{a_0}^{a} \frac{(J - K^2/E)}{(w - a)} (1.76 - 0.76 \, a/w) da$$

Then the procedure of the previous sections was followed in detail, using for the CT specimen of $W = 508$ mm (20 in.) the above formula. The *EPRI Handbook* was used to obtain the needed calibration functions.

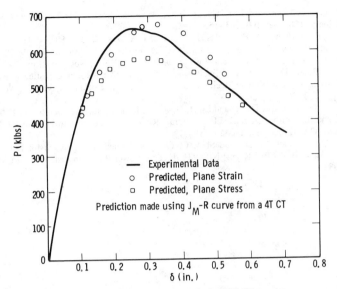

FIG. 1—*Load-displacement record of a 10T CT specimen.*

The predicted as well as the experimentally determined *P-v* records are shown in Fig. 1. The answer to the foregoing questions 3 and 4 is provided by measuring directly the slope of the unloading part of the *P-v* record $-dP/dv$ [3]. This value represents the minimum stiffness in parallel needed to prevent instability, while its inverse $(-dP/dv)^{-1}$ represents the minimum compliance in series needed to cause instability:

$$(a) \qquad P_{max} = 20\ 014\ \text{kN}\ (655.2\ \text{klb})$$

$$(b) \qquad v_{inst} = 8.13\ \text{mm}\ (0.32\ \text{in.})$$

$$(c) \qquad C_{CR} = 5.71\ 10^{-3}\ \text{mm/kN}\ (10^{-3}\ \text{in./klb})$$

$$(d) \qquad K_{min} = 175\ \text{kN/mm}\ (10^3\ \text{klb/in.})$$

References

[1] Ernst, H. A., "Material Resistance and Instability Beyond *J* Controlled Crack Growth," *Elastic-Plastic Fracture: Second Symposium, Volume I—Inelastic Crack Analysis, ASTM STP 803,* C. F. Shih and J. P. Gudas, Eds., American Society for Testing and Materials, Philadelphia, 1983, pp. I-191–I-213.
[2] Ernst, H. A. and Landes, J. D., "Elastic-Plastic Fracture Mechanics Methodology Using the Modified *J*, *J$_M$* Resistance Curve Approach," to be published in the *Journal of Pressure Vessel Technology,* 1985.
[3] Ernst, H. A., "Some Salient Features of the Tearing Instability Theory," Elastic-Plastic Fracture: *Second Symposium Volume II—Fracture Resistance Curves and Engineering Applications, ASTM*

STP 803, C. F. Shih and J. P. Gudas, Eds., American Society for Testing and Materials, Philadelphia, 1983, pp. II-133–II-155.

[4] Albrecht, P., Andrews, W. R., Gudas, J. P., Joyce, J. A., Loss, F. J., McCabe, D. E., Schmidt, D. W., and Van Der Sluys, W. A., "Tentative Test Procedure for Determining the Plain-Strain J-R Curve," *Journal of Testing and Evaluation*, Vol. 10, No. 6, Nov. 1982, pp. 245–251.

[5] Gudas, J. P. and Davis, D. A., "Evaluation of the Tentative J-R Curve Testing Procedure by Round Robin Tests of HY 130 Steel," *Journal of Testing and Evaluation*, Vol. 10, No. 6, Nov. 1982, pp. 252–262.

[6] Landes, J. D., McCabe, D. E., and Ernst, H. A., "Elastic Plastic Methodology to Establish R-Curves and Instability Criteria," EPRI Contract RP 1238-2, Electric Power Research Institute, Palo Alto, CA, (Final Report, March 1983).

[7] Kumar, V., German, M. D., and Shih, C. F. "An Engineering Approach for Elastic-Plastic Fracture Analysis," EPRI Report NP 1931, Electric Power Research Institute, Palo Alto, CA, July 1981.

J. C. Newman, Jr.[1]

Prediction of Stable Crack Growth and Instability Using the V_R-Curve Method

REFERENCE: Newman, J. C., Jr., **"Prediction of Stable Crack Growth and Instability Using the V_R-Curve Method,"** *Elastic-Plastic Fracture Mechanics Technology, ASTM STP 896*, J. C. Newman, Jr., and F. J. Loss, Eds., American Society for Testing and Materials, Philadelphia, 1985, pp. 139–166.

ABSTRACT: This paper presents a methodology for predicting stable crack growth and instability of cracked structural components from results of laboratory tests on metallic materials under plane-stress conditions. The methodology is based on the displacement (V_R) at the tip of a stably tearing crack. Basically, the V_R-curve method is a resistance curve approach, such as K_R and J_R, except that the "crack drive" is written in terms of crack-tip displacement instead of K or J. The relationship between crack-tip-opening displacement, crack length, specimen type, and tensile properties is derived from the Dugdale model for the cracked structure of interest.

This report describes the laboratory test procedure and calculations used to obtain the V_R resistance curve from fracture tests of compact or of middle-crack tension (formally center-crack) specimens. The analysis procedure used to predict stable crack growth and instability of any through-the-thickness crack configuration made of the same material and thickness, and tested under the same environmental conditions, is presented. The various limitations of the present V_R-curve method are given. Four example calculations and predictions are shown.

KEY WORDS: fracture strength, test methods, toughness, cracks, fracture properties, elastic properties, plastic properties

Nomenclature

a Crack length, m

a_0 Initial crack length, m

B Specimen thickness, m

C Constant in V_R equation (Eq 5)

d Crack length plus tensile plastic zone ($a + \rho$), m

E Modulus of elasticity, N/m^2

F_i, H_i Boundary-correction factors on stress intensity and on displacement

J_R Material crack-growth resistance in terms of J, N/m

K Stress-intensity factor, $N/m^{-3/2}$

K_A Applied (crack-drive) stress-intensity factor, $N/m^{-3/2}$

K_R Material crack-growth resistance in terms of K, $N/m^{-3/2}$

---

[1] Senior scientist, NASA Langley Research Center, Hampton, VA 23665.

M Number of data points

n Constant in V_R equation (Eq 5)

P Applied load, N

P_f Failure load, N

S Remote uniform stress, N/m^2

V_a One-half crack-tip-opening displacement (CTOD) at current crack tip, m

V_A Applied (crack-drive) one-half CTOD, m

V_c Critical one-half CTOD at current crack tip, m

V_i Constant in V_R equation (Eq 5), m

V_{pw} Difference in one-half CTOD from stationary and growing crack, m

V_R Material crack-growth resistance in terms of V_a, m

w One-half width or width of specimen (see Fig. 1), m

Y Backface yield correction on CTOD for compact specimen

β Plastic zone to width (ρ/w) ratio

γ Plastic zone to "fictitious" crack length (ρ/d) ratio

Δa Physical crack extension, m

λ "Fictitious" crack length to width (d/w) ratio

ξ Crack length to width (a/w) ratio

ρ Length of tensile plastic zone, m

ρ_0 Plastic-zone size at incipient yield at Point A for compact specimen, m

σ_0 Flow stress of material ($\sigma_{ys} + \sigma_u$)/2, N/m^2

σ_{ys} Yield stress (0.2% offset), N/m^2

σ_u Ultimate tensile strength, N/m^2

ϕ Unit load function for CTOD due to applied loading

ψ Unit stress function for CTOD due to flow stress

In damage-tolerant and structural-integrity analyses, the residual strength of a flawed component must be evaluated. The residual strength is the maximum load-carrying capacity of the flawed structure. For brittle materials, linear-elastic fracture mechanics (LEFM) concepts, such as K_{Ic} (plane-strain fracture toughness), are used. But for materials that exhibit large amounts of plasticity at the crack tip and stable crack growth prior to failure, LEFM concepts are not applicable. For these materials, methods which account for plasticity and stable crack growth should be used.

The Dugdale model [1] is a very simple approach that simulates the effects of plasticity on plastic-zone size and on crack-tip-opening displacements (CTOD) for thin materials. The model does not, however, accurately model yielding under plane-strain conditions. But the model is adopted herein because of its mathematical simplicity. The Dugdale model concept has been used with middle-crack tension (formerly center-crack) specimens in several fracture analyses [2–4] for cracked metallic materials. These fracture analyses have had varying degrees of success. These approaches, however, have lacked the versatility of the resistance-curve concepts in accounting from both plasticity and stable crack growth. Wnuk

[5] has developed a stable crack growth analysis based on modifications to the Dugdale model. His analysis, however, has lacked experimental confirmation.

The resistance curve methods are used to characterize the resistance to fracture during slow-stable crack extension in metallic materials. Unlike brittle fracture, which is characterized by a single value of fracture toughness (K_{Ic}), the resistance curve provides a toughness record as a crack is driven stably into the plastic zone caused by increasing applied loads. The resistance curve, properly calculated, is unique for the material thickness of interest. It is independent of crack length, specimen width, and specimen type for through-the-thickness cracks. It is dependent, however, upon specimen thickness, temperature, environmental conditions, and strain rate. These resistance curves are used to predict stable crack growth and instability (maximum) load for cracked structural components.

The K_R resistance curve method is based on stress-intensity factor analyses and, consequently, applies only for cracks with small to moderate plasticity. The J_R resistance curve method, on the other hand, is able to account for large amounts of plasticity at the crack tip, but accurate relations between J_R, crack length, specimen type, and tensile properties sometimes require an elastic-plastic finite-element analysis (see Ref 6, for example).

The V_R resistance curve method, developed herein, is quite similar to the J_R resistance curve method except that the "crack drive" is written in terms of crack-tip displacement instead of the J-integral. A relationship between the crack-tip-opening displacement (V_R), crack length, specimen type, and tensile properties has been derived from the Dugdale model. The Dugdale model solutions are easily obtained from superposition of two elastic crack problems. Consequently, the V_R-curve method can be applied to any crack configuration for which these two elastic solutions have been obtained.

This report presents the V_R-curve method in the form of a recommended guide. The report describes the laboratory test procedure and calculations used to obtain the V_R resistance curve from fracture tests of compact or of middle-crack tension specimens made of metallic materials under plane-stress conditions. For structural application, the procedure used to predict stable crack growth and instability of any through-the-thickness crack configuration made of the same material and thickness, and tested under the same environmental conditions, is presented. Various limitations of the method are given. Four example calculations and predictions are shown.

Scope

This report covers the determination of the resistance to fracture of metallic materials under plane-stress conditions. The material resistance is written in terms of the crack-tip-opening displacement (CTOD), V_R, as a function of physical crack extension. The Dugdale model for the compact and middle-crack tension specimen, as shown in Fig. 1, was used to obtain plastic-zone size and displacement equations that are used to calculate V_R.

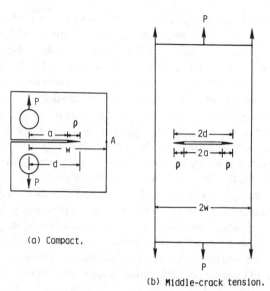

(a) Compact.

(b) Middle-crack tension.

FIG. 1—*Crack configurations analyzed with the Dugdale model.*

Laboratory specimens of standard proportions with constant thickness are required, but specimen width and crack length are variable. The largest, most practical, specimen width is recommended.

Metallic materials that can be tested are not limited by strength, thickness, or toughness. However, various limitations on the method will preclude extremely tough, high strain-hardening materials under plane-strain conditions. Furthermore, the crack must grow under ductile tearing.

For structural application, the Dugdale model solution for the cracked structure of interest must be obtained. The plastic-zone size and "crack-drive" CTOD, V_A, are used in the resistance-curve concept to predict stable crack growth and instability.

Applicable Documents

ASTM Standards:

E 8	Tension Testing of Metallic Materials
E 399	Test for Plane-Strain Fracture Toughness of Metallic Materials
E 561	Test for R-Curve Determination
E 616	Terminology Relating to Fracture Testing

Summary of Method

The V_R resistance curve is determined from load-crack extension measurements or failure load data on compact or middle-crack tension specimens. The specimens

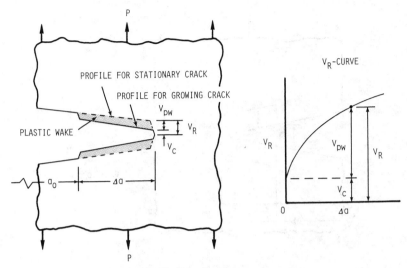

FIG. 2—*The V_R resistance curve concept.*

must be of constant thickness. Various limitations on the V_R-curve method must be met so that the resistance curve will be unique, that is, be independent of crack length, specimen width, and specimen type.

The V_R resistance curve can be used to predict stable crack growth and instability of cracked structures provided a Dugdale model solution is available. These solutions can be obtained by superposition of two elastic crack problems.

Significance

The V_R resistance curve method, developed herein, is quite similar to the K_R or J_R resistance curve methods except that the "crack drive" is written in terms of crack-tip displacements instead of K or J. Figure 2 illustrates the definition of V_R. A cracked plate with an initial crack length, a_0, is subjected to an applied load P which causes the crack to stably tear into the plastic zone and leave behind the plastic wake (shaded region). The critical crack-tip-opening displacement (V_c) has been shown to be constant during stable crack growth both experimentally [7] and analytically [8,9]. The plastic wake is the difference in the crack surface profile for a stationary crack and for a growing crack. V_{pw} is the plastic-wake displacement. The displacement V_R is defined as the sum of $V_c + V_{pw}$. In the V_R hypothesis, it is assumed that V_R is uniquely related to the amount of crack extension, Δa.

The relationship between crack-tip-opening displacement (V_R), crack length, specimen type, and tensile properties used herein is derived from the Dugdale model, as shown in Fig. 3. (For the Dugdale model, the J-integral is equal to a constant times the crack-tip-opening displacement.) Dugdale model solutions are easily obtained from superposition of two elastic crack problems. Consequently,

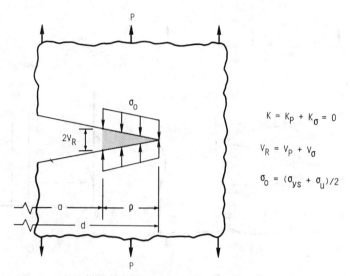

$$K = K_P + K_\sigma = 0$$

$$V_R = V_P + V_\sigma$$

$$\sigma_0 = (\sigma_{ys} + \sigma_u)/2$$

FIG. 3—*Crack-growth resistance, V_R, calculated from the Dugdale model.*

the V_R-curve method can be applied to any crack configuration for which these two elastic solutions have been obtained.

The V_R-curve concept presented here is different from the δ_R-curve concept presented in Ref *10*. In the δ_R-curve concept, the crack-tip-opening displacement is measured (or calculated) at a location near the initial crack tip (prior to load application), whereas, in the V_R-curve concept, the crack-tip-opening displacement is at the current crack tip. However, the V_R displacement cannot be measured because it involves the difference in displacements from two crack states (stationary and growing).

Laboratory Test

The objective of the laboratory test is to develop the material resistance to stable crack growth in terms of the crack-tip-opening displacement (CTOD) at the current crack tip, that is, V_R, against physical crack extension (Δa). The test procedure is identical to various sections in ASTM E 561, insofar as the determination of load against physical crack-extension data is concerned. Some of these sections will be repeated here for completeness. Although the methodology applies to many crack configurations, detailed information on plastic-zone size and CTOD is given only for compact and middle-crack tension specimens (Appendix I). The following will describe the test specimen design and preparation, test procedure, calculation and interpretation of results, and results to be reported.

Test Specimen Design and Preparation

Compact Specimen—The specimen design, preparation, grips, and fixtures described in ASTM E 399 are recommended for V_R-curve testing of the compact specimen shown in Fig. 1a. There are, however, no restrictions on specimen thickness. For sheet specimens, the portion of the specimen arms and backface which are in compression should be restrained from buckling (see ASTM E 561 for details).

All specimens must be fatigue precracked. In precracking, the minimum-to-maximum load ratio can be chosen through experience, but a ratio of 0.1 is commonly used. The ratio of maximum stress-intensity factor of the fatigue cycle to the modulus of elasticity (K_{max}/E) shall not exceed 0.0003 m$^{1/2}$ (0.002 in.$^{1/2}$).

Middle-Crack Tension Specimen—The specimen design, preparation, grips, and fixtures described in ASTM E 561 are recommended for V$_R$-curve testing of the middle-crack specimen shown in Fig. 1b. Again, for sheet specimens, the portion of the specimen in compression (along crack surfaces) should be restrained from buckling (see ASTM E 561 for details).

The fatigue precracking procedure described for compact specimens should also be used for middle-crack specimens.

Test Procedure

The V$_R$-curve can be developed by using two different methods. In the first, the load against physical crack-extension data is measured from either unloading compliance or visual observations. In the second method, the V$_R$-curve is determined from failure load data, similar to that proposed in Ref *11*. The test procedures that are common to both methods will be given first and then the procedures used for the two methods will be given separately.

Measure specimen width, *w*, to ±1% of *w*. The specimen thickness, *B,* is to be measured to ±1% of *B* at three locations near the crack plane. Tests must be conducted at constant thickness.

Replicate V$_R$-curves can be expected to vary as do other mechanical properties. At least three tests should be conducted at each crack length and specimen width. Because the extent of the V$_R$-curve is an increasing function of specimen width, the largest, most practical specimen width should be tested. At least nine tests should be used to obtain data for the resistance curve. Besides the largest specimen to be tested, the other specimens are suggested to be sized at one-half width and one-quarter width of the largest specimen. This will develop instability points at other locations along the resistance curve.

Load-Against-Crack-Extension Method—Apply load to the compact or middle-crack specimens incrementally, allowing time between load increments for the crack to stabilize before measuring load and crack length. Cracks stabilize in most materials within seconds of stopping the load. However, when stopping near an instability condition, the crack may take several minutes to stabilize, depending upon the stiffness of the machine. Measure the physical crack length to 0.2 mm (0.01 in.) at each load increment using suitable measuring devices. Physical crack length can also be measured with compliance techniques by partial unloading of the specimen after each increment as described in ASTM E 561.

Tests may be conducted under load-control or displacement-control conditions.

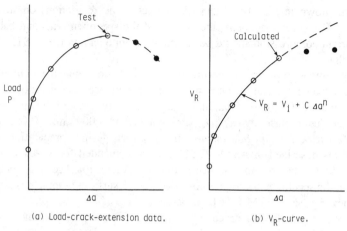

(a) Load-crack-extension data. (b) V_R-curve.

FIG. 4—*Determination of the V_R-curve from load-against-crack-extension data.*

Under displacement-control, however, crack-extension data beyond maximum load should be recorded but should not be used in developing the V_R-curve. The resistance curve, at least in terms of V_R, may not be a material property beyond maximum load. (This behavior will be discussed and experimentally shown later.)

Typical load-crack-extension data are shown in Fig. 4a. Each data point up to and including the maximum load point (open symbols) is used to calculate a point on the V_R-curve, Fig. 4b. Crack-extension data beyond maximum load (solid symbols) will not necessarily lie on the V_R-curve. The V_R equation is then fitted to the valid V_R-Δa data (open symbols) using the least-squares procedure outlined in Appendix II.

Failure Load Method—To determine the V_R-curve by using the failure load method, fracture tests must be conducted at different specimen widths or different crack lengths or both. The extent of the V_R-curve is directly related to specimen width. Small-width specimens become unstable at low points on the resistance curve, whereas large-width specimens become unstable at high points. It is recommended that specimen widths be chosen to have the widest range possible with a constant a_0/w ratio. For the compact specimen, $a_0/w = 0.4$ to 0.6 is recommended; for the middle-crack specimen, $a_0/w = 0.3$ to 0.4 is recommended. Tests may be conducted under load-control or displacement-control. The maximum load should be recorded from a load-displacement (crack mouth or ram) record or from a peak-load meter.

Figure 5a shows some typical failure load data on different-width specimens at constant a_0/w. Each point should be the average failure load on several tests (at least three). Each failure load point (symbol) is then used to calculate a point on the V_R-curve, Fig. 5b. The procedure used to calculate V_R is given in Appendix III. The V_R equation is then fitted to the V_R-Δa data using the procedure in Appendix II.

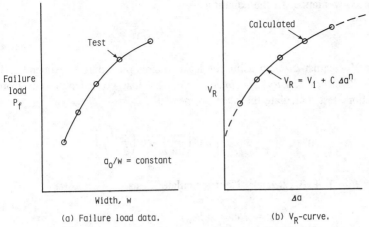

(a) Failure load data.   (b) V_R-curve.

FIG. 5—*Determination of the V_R-curve from failure load data.*

Calculation and Interpretation of Results

Many crack configurations could be used to develop the V_R-curve. However, the compact and middle-crack tension specimens, being the most widely tested, were selected as the laboratory specimens.

Compact Specimen—From the load-crack-length data determined from either of the laboratory test methods, calculate the plastic-zone size as

$$\rho = \frac{\pi}{8}\left(\frac{K}{\sigma_0}\right)^2 F_0 \tag{1}$$

for the compact specimen [12]. The function F_0 is defined in Appendix I. Next, calculate V_R as

$$V_R = P\phi(a,w,B,\rho) - \sigma_0\psi(a,w,B,\rho) \tag{2}$$

where

$$\phi = \frac{4}{BE}\sqrt{\frac{\beta}{2\pi}}\,F_1 F_2 Y \tag{3}$$

and

$$\psi = \frac{w}{2\pi E}\sqrt{\beta}\,H_1 H_2 Y \tag{4}$$

for each P-Δa data point. The functions β, F_1, F_2, H_1, H_2, and Y are defined in Appendix I. See the section on "Limitations of the Method" and Appendix I for limitations on Eqs 1 and 2.

For convenience, fit the equation

$$V_R = V_i + C(\Delta a)^n \tag{5}$$

to the V_R-against-Δa data using the least-squares procedure in Appendix II.

Middle-Crack Tension Specimen—Using the load-crack-length data from the laboratory test, calculate the plastic-zone size as

$$\rho = a\left\{\frac{2w}{\pi a} \sin^{-1}\left[\sin\left(\frac{\pi a}{2w}\right) \sec\left(\frac{\pi Sf}{2\sigma_0}\right)\right] - 1\right\} \tag{6}$$

where $f = 1 + 0.22(a/w)^2$. Next, calculate V_R as

$$V_R = P\phi(a,w,B,\rho) - \sigma_0\psi(a,w,B,\rho) \tag{7}$$

where

$$\phi = \frac{2}{wBE} \sqrt{d^2 - a^2}\, F \tag{8}$$

and

$$\psi = \frac{2}{\pi E}\left\{\left[\pi - 2\sin^{-1}\left(\frac{a}{d}\right)\right] \sqrt{d^2 - a^2}\right.$$
$$\left. - 2a\cosh^{-1}\left(\frac{d^2 + a^2}{2ad}\right)\right\}\frac{H}{fg} \tag{9}$$

for each P-Δa data point. The functions F, g, and H are defined in Appendix I. See the section on "Limitations of the Method" and Appendix I for limitations on Eqs 6 and 7.

Again, fit the V_R equation (Eq 5) to the V_R-against-Δa data for the middle-crack tension specimens.

Report Results

1. Type and size of specimen tested.
2. Crack direction in the material (see ASTM E 399).
3. Tensile properties (σ_{ys}, σ_u, and E).
4. Specimen thickness (B).
5. Test temperature.
6. Three constants in V_R equation (V_i, C, and n).

Structural Application

In the following, the analysis required to calculate the "crack-drive" displacement (V_A), the procedure used to predict stable crack growth and instability, and the results to be reported are presented.

Analysis Required

To apply the present method, a Dugdale model solution is needed for the crack configuration of interest. Plastic-zone size and crack-tip-opening displacement results are needed for various crack lengths. These quantities expressed in equation form would be helpful. The primary advantage in using this model is that plastic-zone size and displacements are obtained by superposition of two elastic problems. These two elastic problems, a through-the-thickness crack of length *d* in the configuration of interest subjected first to the required loading and second to a uniform stress (σ_0) acting over the plastic-zone length ρ, as shown in Fig. 3, must be obtained. Various elastic analyses or handbook solutions may be used to obtain the required solutions. Some elastic analyses that may be used are the boundary-collocation methods, the finite-element methods, the boundary-integral method, and the body-force method.

The plastic-zone size for a crack in the structure of interest may be determined by requiring that the finiteness condition of Dugdale be satisfied [1]. This condition states that the stress-intensity factor at the tip of the plastic zone is zero ($a + \rho$) and is given by

$$K_P + K_{\sigma_0} = 0 \qquad (10)$$

where K_P is the stress-intensity factor due to the applied load and K_{σ_0} is the stress-intensity factor due to the flow stress. The plastic-zone size is determined by satisfying Eq 10. Once the plastic-zone size has been determined, then the displacement at the current crack tip (a) must be calculated (see Fig. 3). The crack-drive displacement is calculated as

$$V_A = V_P + V_{\sigma_0} \qquad (11)$$

where V_P is the displacement due to the applied load and V_{σ_0} is the displacement due to the flow stress.

For materials that fracture under small-scale yield conditions, some simple equations for plastic-zone size and CTOD can be developed. Usually, if ρ/a and $\rho/(w - a)$ are less than 0.1, then small-scale yield conditions exist. The advantage in this approach is that ρ and CTOD equations are expressed in terms of the elastic stress-intensity factor. Assuming that the applied load is small, the plastic-zone size is

$$\rho = \frac{\pi}{8}\left(\frac{K_A}{\sigma_0}\right)^2 \qquad (12)$$

where K_A is stress-intensity factor at the current crack length. Similarly, one-half the CTOD is

$$V_A = \frac{K_A^2}{2\sigma_0 E} \qquad (13)$$

Appendix IV gives further details and an example on the use of small-scale yield solutions to predict stable crack growth and fracture.

Procedure and Calculation of Results

The procedure and calculations required to predict stable crack growth and instability under load-control and displacement-control conditions are presented here. Under either condition, the calculations are valid only up to maximum load. Beyond maximum load, under decreasing load conditions, the method has not been evaluated.

The V_R-curve is plotted on a displacement against crack length coordinate system by placing the origin of the curve at the initial crack length, as shown in Fig. 6. If the cracked structure is under load-control, the crack-drive (dash-dot) curve, V_A, is calculated at a fixed load as a function of crack length. At constant load, P_1, P_2, or P_3, the intercept of the V_A-curve and the V_R-curve at Points A, B, and C, respectively, gives the amount of stable crack growth $(a - a_0)$. At load P_f, the V_A-curve becomes tangent to the V_R-curve at Point D. This is the instability point. At each point A, B, C, and D, the limitations of the method must be checked. Calculations should not be made beyond these limitations.

For displacement-control conditions, crack instability may occur beyond maximum load. In the present method, however, calculations should only be made up to maximum load. A plot similar to that shown in Fig. 6 is to be constructed. The crack-drive curves, however, are calculated at constant displacement, so long as the loads at the intercept points A, B, C, and D are monotonically

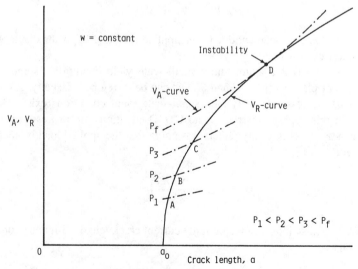

FIG. 6—*Illustration of the V_R-curve concept with "crack-drive" V_A-curves.*

increasing. (Further study is needed to extend the V$_R$-curve concept to predict instabilities beyond maximum load.)

Report Results

The predicted load-against-crack-extension $(P, \Delta a)$ values should be plotted or listed. The instability load should also be given.

Limitations of the Method

The limitations of the present V$_R$-curve method have been grouped into two categories: those that are common to the method and would be applied to both laboratory specimens and structural application, and those that are particular to the laboratory specimens used to obtain the V$_R$-curve.

Common Limitations

1. The crack must grow by stable tearing.
2. The crack-growth mechanism and degree of constraint (state of stress) are the same in laboratory test specimens and structural application. To achieve these conditions, the specimen thickness must be the same; and the crack length (a) and uncracked ligament $(w - a)$ must be greater than the specimen thickness (B).
3. The plastic-zone size

$$\rho = \rho(P, \sigma_0, a, w, B)$$

and displacement at the crack tip

$$V_a = P\phi(a, w, B, \rho) - \sigma_0\psi(a, w, B, \rho)$$

from the Dugdale model for the crack configuration of interest must be available. For the Dugdale model with constant flow stress (σ_0), the ratio of σ_u/σ_{ys} must be less than 1.6. The total crack length plus plastic-zone size cannot be larger than the specimen width. Also, the influence of yielding at locations other than the crack tip on ρ and V_a must be accounted for in the analysis.
4. Failure load predictions must not be carried on further, in crack extension, than the range of data available from laboratory tests. Thus, the largest, most practical specimen width should be used in obtaining the V$_R$-curve.

Particular Limitations for Laboratory Specimens

1. For the compact specimen, see the particular limitations listed for the plastic-zone and CTOD equations in Appendix I. In determining the V$_R$-curve, the crack

length plus plastic zone $(a + \rho)$ must be less than $0.8\ w$. Also, the backface-yield correction (Y) given by

$$Y = 1 + 0.25\left(\frac{\rho}{\rho_0} - 1\right)^4 \left(\frac{w}{a}\right) \tag{14}$$

where

$$\frac{\rho_0}{w - a} = 0.48 - 0.84\,\frac{a}{w} + 0.56\left(\frac{a}{w}\right)^2 \tag{15}$$

must be less than 1.05 (see Appendix I) for $\rho \geq \rho_0$. To predict failure loads on compact specimens, however, the limitation on Y may be relaxed and Y used up to failure.

2. For the middle-crack tension specimen, see the particular limitations listed for the plastic-zone and CTOD equations in Appendix I. In determining the V_R-curve, the crack length plus plastic zone $(a + \rho)$ must be less than $0.85\ w$ and the net-section stress S_n must be less than $0.85\ \sigma_0$. In predicting failure loads on these specimens, the net-section stress limitation may be ignored.

3. Load-crack extension data must not include data with crack-extension values beyond maximum load in displacement-control tests.

Example Calculations and Predictions

In this section, three example crack problems are considered. These example problems will demonstrate how the V_R-curve is obtained from laboratory tests and how the V_R-curve concept is used to predict stable crack growth and instability on other cracked specimens made of the same material and thickness. First, load-crack extension data $(P, \Delta a)$ on compact specimens made of 2024-T351 aluminum alloy are used to obtain the V_R-curve. This curve is then used to predict failure of a middle-crack tension specimen. Second, failure load data (P_f, a_0) on compact specimens made of 7075-T651 aluminum alloy are used to obtain the V_R-curve. Again, this curve is used to predict failure loads on two different-size middle-crack tension specimens. Last, the V_R-curve obtained for the 7075-T651 material is used to predict failure load on a structurally-configured specimen. The structurally configured specimen, containing three circular holes with a crack emanating from one of the holes, was subjected to tensile loading [9].

Problem 1

The first example is to predict the failure load on a middle-crack tension specimen from results of fracture tests conducted on compact specimens made of 2024-T351 aluminum alloy plate $(B = 12.7\ \text{mm})$. Various-size compact specimens $(w = 51, 102, \text{and } 203\ \text{mm})$ were precracked according to ASTM E 399 requirements and monotonically pulled to failure under displacement control [13].

At various load levels, the specimens were partially unloaded (10 to 15%) and the unloading compliance was measured. A predetermined relationship between compliance and crack length was then used to obtain the physical crack length and, subsequently, crack extension, Δa.

For each applied load and crack extension value $(P, \Delta a)$, the crack-growth resistance, V_R, was calculated from

$$V_R = P\phi(a,B,w,\rho) - \sigma_0\psi(a,B,w,\rho) \tag{16}$$

Equation 16 is identical to Eq 22 in Appendix I with $V_a Y = V_R$ and ρ is calculated from Eq 1. The values of V_R and corresponding Δa are plotted in Fig. 7a. Solid symbols denote crack extension values beyond maximum load (invalid data, see limitations). The reason for this behavior is not completely understood at present. The V_R equation was fitted to the open symbols by using the least-squares procedure described in Appendix II. Caution must be exercised when using the V_R equation beyond the range of valid data (dashed curve). Larger-width specimens than those shown in Fig. 7a may be used to obtain valid data over a wider range of crack extension than that shown.

Figure 7b demonstrates how the V_R-curve concept is used to predict stable crack growth and instability (failure) load on the middle-crack tension specimen. The specimen had a half-width of 127 mm with an initial crack length of 52.1 mm [14]. The figure shows crack-tip-opening displacement, V_A and V_R, plotted against crack length. As usual for resistance-curve methods, the V_R curve is positioned at a_0. Again the dashed curve is the estimated V_R-curve beyond the range of experimental data. The "crack-driving force" V_A-curves are shown as the dash-dot curves and were calculated from

$$V_A = P\phi(a,B,w,\rho) - \sigma_0\psi(a,B,w,\rho) \tag{17}$$

at constant load. Equation 17 is identical to Eq 22 in Appendix I with $V_a Y = V_A$ and ρ is calculated from Eq 1. The intercept of the V_A and V_R curve gives the stable crack length (a) and, subsequently, crack extension $(a - a_0)$ at the corresponding load. At this point, a further increase in load is required to extend the crack. When the V_A-curve becomes tangent to the V_R-curve at $P_f = 605$ kN, crack growth is unstable and the specimen fails. The predicted failure load was about 5.5% higher than the experimental failure load. No measurement of crack extension was made on the middle-crack specimen.

Problem 2

The second example is to predict the failure load on two different-size middle-crack tension specimens made of 7075-T651 aluminum alloy plate $(B = 12.7$ mm) from results of tests conducted on compact specimens. Again, various-size

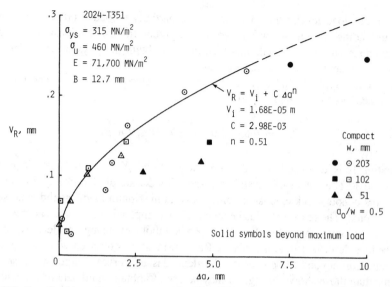

FIG. 7a—Crack-growth resistance curve for 2024-T351 aluminum alloy compact specimens.

compact specimens were tested [13,14] with an identical test procedure as described in Problem 1.

In this example, however, the V_R-curve was determined in two ways. The resistance curve was determined from either load-crack extension data ($P, \Delta a$) or from failure load data (P_f, a_0). Figure 8a shows V_R plotted against Δa for the aluminum alloy compact specimens. Again, Eq 16 was used to calculate V_R (open symbols) from the load-crack extension data. The solid symbols denote the instability points estimated from the failure load data (as described in Appendix III). With the exception of the last two data points at Δa of about 13 mm, the open and solid symbols agreed well. The solid and dashed curves show the V_R equation that was fitted to the respective data. In the following, the V_R-curve determined from the failure load data (dashed curve) will be used. The dashed curve is believed to be more accurate than the solid curve because it is based on the average of five or six tests whereas the solid curve is based on measurements made on only one specimen for each specimen size.

Figure 8b shows how the V_R-curve concept is used to predict failure load on two different size middle-crack tension specimens [14]. Again, V_A and V_R are plotted against crack length. The load which makes the crack-drive (V_A) curve tangent to the corresponding V_R-curve is the failure load for that specimen. The predicted failure loads were 7% to 10% lower than the experimental failure loads. Again, no measurement of crack extension was made on the middle-crack specimen.

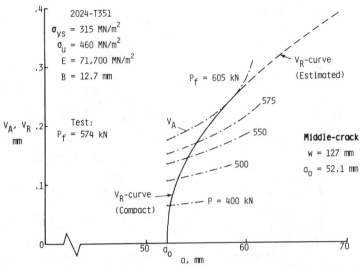

FIG. 7b—*Application of the V_R-curve concept to predict stable crack growth and failure load on middle-crack tension specimen made of 2024-T351 aluminum alloy.*

Problem 3

In the last example, the V_R-curve determined from failure load data on 7075-T651 aluminum alloy compact specimens (Fig. 8a) is used to predict failure load on a three-hole-crack tension specimen [9]. Because the Dugdale model solution for this cracked specimen was unknown, a small-scale yield solution was used, as described in Appendix IV. A small-scale yield solution is justified because the plastic-zone size at failure for the 7075-T651 material is small compared with crack length.

The insert in Fig. 9 shows the three-hole-crack specimen. Further details on this specimen are given in Refs 9 and 14. This specimen had an initial crack length, a_0, of 25.5 mm. The figure shows V_A and V_R plotted against crack length. First, the V_R-curve was positioned at the initial crack length. The dashed curve shows the extension of the V_R-curve beyond the range of valid data. Note that in this specimen, crack extension prior to instability is expected to exceed 50 mm because of the "crack-drive" produced by the two large holes. The lower dashed-dot curve shows the crack-drive, V_A, for this specimen at an applied load of 400 kN. The crack-drive was computed from the small-scale yield solution as

$$V_A = \frac{K_A^2}{2\sigma_0 E} \tag{18}$$

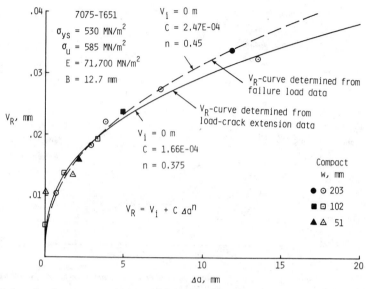

FIG. 8a—*Crack-growth resistance curves for 7075-T651 aluminum alloy compact specimens.*

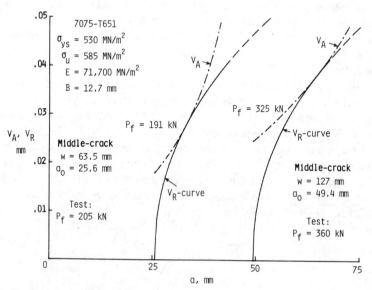

FIG. 8b—*Application of the V_R-curve concept to predict failure loads on middle-crack tension specimens made of 7075-T651 aluminum alloy.*

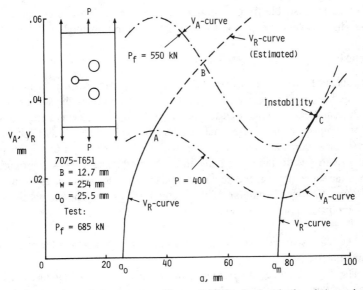

FIG. 9—*Application of the V$_R$-curve concept to predict failure load on the three-hole-crack tension specimen made of 7075-T651 aluminum alloy.*

where K_A is the stress-intensity factor solution from Ref *14*. At $P = 400$ kN, the crack was predicted to grow about 10 mm (Point A). At $P = 550$ kN, the crack would have grown about 30 mm (Point B), far beyond the range of valid data. At 850 kN, the V_A-curve (not shown) would have been tangent to the V_R-curve (dashed curve) at a crack length of about 80 mm. This failure load was about 25% higher than the test load (685 kN). This discrepancy may be due to a combination of two approximations. First, the effects of yielding at the edge of the small hole was neglected. Yielding at the small hole would have resulted in a higher value of V_A at a given crack length and load and, thus, would have given a lower predicted failure load than 850 kN. Second, the range of valid data was severely limited. The shape of the V_R-curve beyond 15 mm's of crack extension (dashed curve) may be in error.

Because the range of valid data was severely limited, a lower-bound estimate for failure load was made. Noting that the crack should stably tear until the minimum in the V_A-curve is reached, the V_R-curve is then positioned at a_m (crack length at minimum V_A). The value of a_m is about 76 mm. (This procedure neglects any effect of prior stable crack growth history on the predicted instability load.) Again, the load that makes the crack-drive V_A tangent to the V_R-curve (Point C) is the lower-bound instability load. The predicted lower-bound failure load was 550 kN. This load was 20% lower than the actual failure load of 685 kN.

Further study is needed on how to predict failure of cracked structures, such as the three-hole-crack specimen, from the limited range of crack-extension data that would be available from laboratory tests of reasonable-size compact specimens. This problem is common to all resistance-curve methods.

Conclusions

A resistance curve method based on crack-tip-opening displacement, the V_R-curve method, was developed herein. The report describes the laboratory test procedure and calculations used to obtain the V_R resistance curve from fracture tests of compact or of middle-crack tension (formally center-crack) specimens. Analysis procedures are also given that will enable the analyst to predict stable crack growth and instability of any through-the-thickness crack configuration made of the same material and thickness, and tested under the same environmental conditions. Various limitations of the method are given. The results presented in this report support the following conclusions.

1. The V_R resistance curve is independent of crack length, specimen width, and specimen type.
2. The V_R resistance curve may be determined from load against crack extension data or from failure load against initial crack length data.
3. The V_R resistance curve concept is based on elastic stress-intensity factors and elastic crack-surface displacements (or the Dugdale model).

APPENDIX I

Plastic-Zone Size and CTOD Equations for Dugdale Model

The plastic-zone size and CTOD equations for the compact and middle-crack tension specimens are presented herein. These equations were used to calculate V_R and V_A for these specimens. Boundary-collocation analyses [15,16] were used to solve the boundary-value problem of the Dugdale model [1] for these specimen types. Equations were then fitted to the boundary-collocation results. For the compact specimen, an elastic-plastic finite-element analysis of the strip-yield model was also conducted to determine the loads required to yield the backface in compression (Point A in Fig. 1a) and to determine its influence on CTOD.

Compact Specimen

The Dugdale model for the compact specimen, Fig. 1a, requires that the "finiteness" condition of Dugdale be satisfied. This condition states that the K at the tip of the plastic zone (at $d = a + \rho$) is zero. From this condition, the plastic-zone size (ρ) was calculated from the boundary-collocation analysis for various a/w and $P/(w\sigma_0 B)$ ratios. An equation was then fitted to these results and was

$$\rho = \frac{\pi w}{8} \left(\frac{PF}{w\sigma_0 B} \right)^2 F_0 = \frac{\pi}{8} \left(\frac{K}{\sigma_0} \right)^2 F_0 \qquad (19)$$

where

$$F = (2 + \xi)(0.886 + 4.64\xi - 13.32\xi^2 + 14.72\xi^3 - 5.6\xi^4)/(1 - \xi)^{3/2}$$

$$F_0 = 1 + C_1\left(\frac{P}{w\sigma_0 B}\right) + C_2\left(\frac{P}{w\sigma_0 B}\right)^2$$

$$C_1 = -2.687(1 - \xi) + 0.167/\xi^2$$

$$C_2 = -2.48 - 0.039/(1 - \xi)^6$$

and $\xi = a/w$. Equation 19 is within 1% of collocation results [12] for $0.3 \leq (a + \rho)/w \leq 0.8$ and $\rho/(w - a) \leq 0.5$.

The CTOD ($2V_a$) for the compact specimen was calculated by adding the displacement at the tip of the physical crack length (a) due to the pin load (P) and due to the uniform stress (σ_0) acting over the plastic-zone length. Again, an equation was fitted to these results and was

$$V_a = \frac{4P}{BE}\sqrt{\frac{\beta}{2\pi}} F_1 F_2 - \frac{w\sigma_0}{2\pi E}\sqrt{\beta} H_1 H_2 \qquad (20)$$

where

$$F_1 = (2 + \lambda)(0.886 + 4.64\lambda - 13.32\lambda^2 + 14.72\lambda^3 - 5.6\lambda^4)/(1 - \lambda)^{3/2}$$

$$F_2 = 1 + B_1\gamma + B_2\gamma^2$$

$$B_1 = -1.25 + 9.76\lambda - 20.15\lambda^2 + 16.62\lambda^3$$

$$B_2 = 0.64 - 4.34\lambda + 10.24\lambda^2 - 8.46\lambda^3$$

$$H_1 = \{[2\beta(1 + A_1 + A_2) + (1 - \lambda)(5 + A_1 - 3A_2)]\sqrt{\beta^2 + (1 - \lambda)\beta}$$

$$+ (1 - \lambda)^2(3 - A_1 + 3A_2)[\ell n (\sqrt{\beta} + \sqrt{\beta + 1 - \lambda}) - \ell n \sqrt{1 - \lambda}]\}/(1 - \lambda)^{3/2}$$

$$A_1 = 3.57 + 12.5(1 - \lambda)^8$$

$$A_2 = 5.1 - 15.32\lambda + 16.58\lambda^2 - 5.97\lambda^3$$

$$H_2 = 1 + D_1\gamma + D_2\gamma^2$$

$$D_1 = 0.666 + 0.796\lambda^2 + 12.36\lambda^4$$

$$D_2 = 0.084 + 2.62\lambda^2 - 14.08\lambda^4$$

$\beta = \rho/w$, $\lambda = (a + \rho)/w$, and $\gamma = \rho/(a + \rho)$. Equation 20 is within about 1.5% of collocation results [12] for $0.3 \leq \lambda \leq 0.8$, $\gamma \leq 1 - 0.2/\lambda$, and $\rho/(w - a) \leq 0.5$.

In the compact specimen, the material at Point A in Fig. 1a is in compression. At a certain load, this material will yield in compression. From a finite-element-strip-yield analysis for an elastic-perfectly plastic material, the load that causes incipient yielding at Point A for various a/w ratios was calculated. The corresponding crack-tip plastic-zone size (ρ_0) at incipient yield at Point A is given by

$$\frac{\rho_0}{w - a} = 0.48 - 0.84 \frac{a}{w} + 0.56\left(\frac{a}{w}\right)^2 \tag{21}$$

When ρ is less than ρ_0, the material at Point A is elastic.

To account for the influence of backface yielding on CTOD, an approximate equation was developed from the results of the finite-element-strip-yield analysis. The crack-tip-opening displacement equation which accounts for backface yielding (BFY) is

$$(V_a)_{\text{BFY}} = V_a Y \tag{22}$$

where V_a is given by Eq 20 and

$$Y = 1 + 0.25\left(\frac{\rho}{\rho_0} - 1\right)^4 \left(\frac{w}{a}\right) \tag{23}$$

for $\rho \geqslant \rho_0$. Y is equal to unity for $\rho < \rho_0$. The ρ_0 value is the crack-tip plastic-zone size at incipient yielding at Point A and is given by Eq 21.

Middle-Crack Tension Specimen

The plastic-zone size (ρ) for a crack in a finite-width specimen (Fig. 1b) was, again, determined by requiring that the finiteness condition of Dugdale be satisfied. From a boundary-collocation analysis, the plastic-zone size was calculated for various a/w and S/σ_0 ratios. An equation was then fitted to these results. The equation selected was similar to an equation derived by Smith [17] from an infinite-periodic array of Dugdale model cracks. The equation is

$$\rho = a\left\{\frac{2w}{\pi a} \sin^{-1}\left[\sin\left(\frac{\pi a}{2w}\right)\sec\left(\frac{\pi S}{2\sigma_0}f\right)\right] - 1\right\} \tag{24}$$

where

$$f = 1 + 0.22\left(\frac{a}{w}\right)^2 \tag{25}$$

Equation 24 is within 1% of the collocation results for $(a + \rho)/w \leqslant 0.85$.

The CTOD ($2V_a$) for the middle-crack tension specimen was calculated by adding the displacement at the tip of the physical crack length (a) due to the remote uniform stress (S) and due to the uniform stress (σ_0) acting over the plastic-zone length. The crack-tip opening displacement is

$$V_a = V_s + V_\sigma \tag{26}$$

The displacement due to the remote stress is

$$V_s = \frac{2S}{E} \sqrt{d^2 - a^2} F = \frac{2P}{BwE} \sqrt{d^2 - a^2} F \qquad (27)$$

where

$$F = \sqrt{\sec\left(\frac{\pi d}{2w}\right)} \qquad (28)$$

The displacement due to σ_0 is

$$V_\sigma = -\frac{2\sigma_0}{\pi E} \left\{ \left[\pi - 2 \sin^{-1}\left(\frac{a}{d}\right) \right] \sqrt{d^2 - a^2} - 2a \cosh^{-1}\left(\frac{d^2 + a^2}{2ad}\right) \right\} \frac{H}{fg} \qquad (29)$$

where

$$H = \left[\frac{\pi - 2 \sin^{-1} Q}{\pi - 2 \sin^{-1}(a/d)} \right] \sqrt{\sec\left(\frac{\pi d}{2w}\right)} \qquad (30)$$

$$Q = \sin\left(\frac{\pi a}{2w}\right) \Big/ \sin\left(\frac{\pi d}{2w}\right) \qquad (31)$$

$$g = 1 - 0.86 \left(\frac{d}{w}\right)^{5/2} \left(\frac{a}{d}\right)\left(1 - \frac{a}{d}\right)^{3/4} \qquad (32)$$

and f is given by Eq 25. Equation 26 is within 2% of the boundary-collocation results for $(a + \rho)/w \leq 0.85$.

APPENDIX II

Least-Squares Procedure for the Determination of the V$_R$-Curve Constants

The equation chosen to fit the experimental data on V_R against Δa is

$$V_R = V_i + C(\Delta a)^n \qquad (33)$$

where V_i, C, and n are assumed to be material constants.

For a given set of data (material, thickness, and test conditions), the constants were obtained by minimizing the sum of the squares of the differences between the calculated V_R from Eq 33 and the experimental value. Because Eq 33 cannot be linearized, the value of n was varied from zero to unity in steps of 0.01 and the corresponding values of V_i and C were determined from a least-squares procedure. The value of n which gave the minimum error (and the corresponding values of V_i and C) was used. If V_i was less than zero, V_i was set equal to zero and the values of C and n were determined from a different least-squares procedure.

For simple notation, let Δ and V be the experimental values of Δa and V_R, respectively. In both least-squares procedures, an arbitrary weighting factor of $\Delta a^{1/2}$ was used. This particular weight factor was needed because of difficulties in determining the initiation of crack growth at $\Delta a = 0$. Using this weight factor, displacements at small Δa values will have less weight than displacements at large Δa values. With the value of n assumed, the sum of the squares of the errors was

$$\sum_{m=1}^{M} e_m^2 = \sum_{m=1}^{M} (V_m - V_i - C \Delta_m^n)^2 \Delta_m^{1/2} \tag{34}$$

where M was the number of data points (Δ_m, V_m). Minimizing the sum of the squares of the errors with respect to V_i and C, and solving, the two equations give

$$C = \frac{W_5 - W_3 W_2 / W_1}{W_4 - W_2 W_2 / W_1} \tag{35}$$

and

$$V_i = \frac{W_3 - CW_2}{W_1} \tag{36}$$

where

$$W_1 = \sum_{m=1}^{M} \Delta_m^{1/2} \qquad W_2 = \sum_{m=1}^{M} \Delta_m^{n+1/2}$$

$$W_3 = \sum_{m=1}^{M} V_m \Delta_m^{1/2} \qquad W_4 = \sum_{m=1}^{M} \Delta_m^{2n+1/2}$$

and

$$W_5 = \sum_{m=1}^{M} V_m \Delta_m^{n+1/2}$$

The value of n which gave the minimum sum of the squares of the errors (Eq 34) was used.

As previously mentioned, if the final V_i was negative, then V_i was set equal to zero and the values of C and n were determined from a different procedure. With $V_i = 0$, the sum of the squares of the errors was

$$\sum_{m=1}^{M} e_m^2 = \sum_{m=1}^{M} (\log V_m - \log C - n \log \Delta_m)^2 \Delta_m^{1/2} \tag{37}$$

Again, minimizing the errors with respect to the unknowns, and solving the two equations, gives

$$n = \frac{W_5 - W_3 W_2 / W_1}{W_4 - W_2 W_2 / W_1} \tag{38}$$

and

$$\log C = (W_3 - nW_2)/W_1 \tag{39}$$

where

$$W_1 = \sum_{m=1}^{M} \Delta_m^{1/2} \qquad W_2 = \sum_{m=1}^{M} \Delta_m^{1/2} \log \Delta_m$$

$$W_3 = \sum_{m=1}^{M} \Delta_m^{1/2} \log V_m \qquad W_4 = \sum_{m=1}^{M} \Delta_m^{1/2} (\log \Delta_m)^2$$

and

$$W_5 = \sum_{m=1}^{M} \Delta_m^{1/2} (\log \Delta_m)(\log V_m)$$

APPENDIX III

Determination of the V$_R$-Curve from Failure Load Data

In 1980, Orange [11] presented a method for estimating the crack-extension resistance curve (K$_R$-curve) from residual-strength (maximum load against initial crack length) data for precracked fracture specimens. Although this elaborate mathematical formation could also have been used here to estimate the crack-tip-opening displacement based V$_R$ curve, a simple graphical method, as discussed by Orange, was used herein.

As pointed out by Orange, it is possible to estimate the resistance curve from residual-strength data by using a purely graphical method. The method is demonstrated in Fig. 10 for 7075-T651 aluminum alloy compact specimens. Here the crack-tip opening displacement is plotted against crack extension. Because the average failure loads (five or six

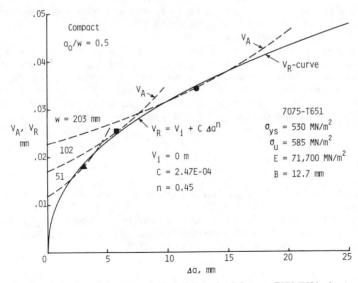

FIG. 10—*Determination of the V$_R$-curve from failure load data on 7075-T651 aluminum alloy compact specimens.*

tests) on three different-size compact specimens ($w = 51$, 102, and 203 mm) were known, crack-driving force (V_A) curves for each specimen size (dashed curve) can be constructed. An estimated R-curve is then drawn from the point ($\Delta a = 0$, $V_R = 0$), monotonically increasing in such a way that it is tangent (or nearly tangent) at some point along each crack driving-force curve. The solid symbol denotes the estimated tangent point for each curve. (Although the graphical method appears simple, in practice it is tedious. If significant data scatter exists, then the selection of the tangent points would be subjective at best.) These three tangent points were then used with the least-squares procedure, described in Appendix II, to determine the V_R-curve (solid curve).

APPENDIX IV

Small-Scale Yield Solutions for Plastic-Zone Size and CTOD

In the following, the small-scale yield solutions for plastic-zone size (ρ) and CTOD are presented. Equations for ρ and CTOD are expressed in terms of the stress-intensity factor. To demonstrate their usefulness, these equations are applied to the fracture of single-edge-crack tension specimens made of 7075-T651 aluminum alloy material.

For materials that fracture under small-scale yield conditions, some simple equations for plastic-zone size and CTOD can be developed. These equations are useful when a Dugdale model solution for ρ and CTOD is unavailable for the cracked structure of interest. These equations may be used in the V_R-curve concept to predict stable crack growth and instability of structures with through-the-thickness cracks. The cracked structure, however, must be of the same material and thickness from which the V_R resistance curve was obtained. The structure must also fail under small-scale yield conditions; that is, the plastic-zone size must be small compared with crack length (a) and must be small compared with the uncracked ligament (such as $w - a$ in the compact and middle-crack specimens). Usually, if ρ/a and $\rho/(w - a)$ are less than 0.1, small-scale yield conditions exist. The primary advantage in this approach is that ρ and CTOD equations are expressed in terms of the elastic stress-intensity factor.

The small-scale yield solution is developed from a crack in an infinite plate. A Dugdale-type yield zone is assumed. The stresses and displacements in a region around the crack tip are assumed to be controlled by an "applied" stress-intensity factor, K_A. The plastic-zone size is calculated from Dugdale's finiteness condition [1]. This condition states that the total stress-intensity factor due to the applied loading and that due to flow stress (σ_0) must be zero. Assuming that the applied load is small, the plastic-zone size is

$$\rho = \frac{\pi}{8}\left(\frac{K_A}{\sigma_0}\right)^2 \tag{40}$$

where K_A is calculated at the current crack length from the stress-intensity factor solution for the cracked structure of interest.

The CTOD at the physical crack tip (a) is obtained by summing the displacement due to the applied stress-intensity factor and that due to σ_0. Again, assuming that the applied load and plastic-zone size are small, one-half the CTOD is

$$V_A = \frac{K_A^2}{2\sigma_0 E} \tag{41}$$

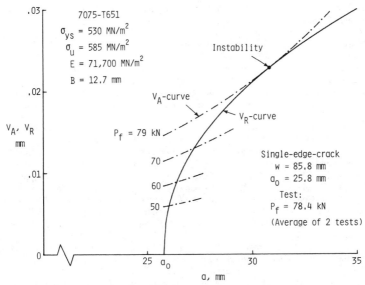

FIG. 11—*Application of the V_R-curve concept to predict stable crack growth and failure load on single-edge-crack tension specimen made of 7075-T651 aluminum alloy.*

V_A is referred to as the "crack-drive" displacement. To calculate ρ and V_A, only the stress-intensity factor solution for the cracked structure of interest is needed.

To demonstrate how the small-scale yield solution is used to predict stable crack growth and failure loads, compact and single-edge-crack tension specimens made of 7075-T651 aluminum alloy material ($B = 12.7$ mm) were tested. The V_R-curve was obtained from the compact specimens (see Fig. 8a or Fig. 10). A least-squares procedure described in Appendix II was used to fit Eq 33 to the experimental data (V_R against Δa). These constants were: $V_i = 0$ m, $C = 2.47 \times 10^{-4}$, and $n = 0.45$. The flow stress (σ_0) was 558 MPa and the modulus of elasticity (E) was 71 700 MPa.

Figure 11 shows the displacements, V_A and V_R, plotted against crack length. The average initial crack length for the single-edge-crack specimen was 25.8 mm. The origin of the V_R-curve is placed at the average initial crack length. Crack-drive curves, shown as dash-dot curves, were calculated from Eq 41 using the applied load as indicated. The stress-intensity factor solution for this specimen was obtained from Tada et al [18]. The intercept of the solid and dash-dot curves gives the amount of crack extension ($a - a_0$) at the corresponding load. The applied load (79 kN) that makes the crack-drive curve tangent to the V_R-curve is the failure (instability) load. This failure load was 0.5% higher than the average experimental failure load (78.4 kN) on two specimens.

References

[1] Dugdale, D. S., "Yielding of Steel Sheets Containing Slits," *Journal of the Mechanics and Physics of Solids,* Vol. 8, 1960, pp. 100–104.

[2] Newman, J. C., Jr., "Fracture of Cracked Plates Under Plane Stress," *Engineering Fracture Mechanics,* Vol. 1, No. 1, 1968, pp. 137–154.

[3] Heald, P. T., Spinks, G. M., and Worthington, P. J., "Post-Yield Fracture Mechanics," *Material Science and Engineering,* Vol. 10, 1972, pp. 129–138.

[4] Larsson, L. H., "Use of EPFM in Design," *Advances in Elasto-Plastic Fracture Mechanics*, Applied Science Publishers, Ltd., Essex, England, 1979, pp. 261–278.

[5] Wnuk, M. P., "Occurrence of Catastropic Fracture in Fully Yielded Components—Stability Analysis," *International Journal of Fracture*, Vol. 15, No. 6, 1979, pp. 553–580.

[6] Kumar, V. and Shih, C. F., "Fully Plastic Crack Solutions, Estimation Scheme, and Stability Analyses for the Compact Specimen," *Fracture Mechanics: Twelfth Conference, ASTM STP 700*, American Society for Testing and Materials, Philadelphia, 1980, pp. 406–438.

[7] Luxmoore, A., Light, M. F., and Evans, W. T., "A Comparison of Energy Release Rates, the J-Integral and Crack Tip Displacements," *International Journal of Fracture*, Vol. 13, 1977, pp. 257–259.

[8] de Koning, A. U., "A Contribution to the Analysis of Slow Stable Crack Growth," National Aerospace Laboratory Report NLR MP 75035 U, The Netherlands, 1975.

[9] Newman, J. C., Jr., "An Elastic Plastic Finite-Element Analysis of Initiation, Stable Crack Growth, and Instability," *Fracture Mechanics: Fifteenth Symposium, ASTM STP 833*, R. J. Sanford, Ed., American Society for Testing and Materials, Philadelphia, 1984, pp. 93–117.

[10] Hellman, D. and Schwalbe, K. H., "Geometry and Size Effects on J_R and δ_R Curves Under Plane Stress Conditions," *Fracture Mechanics: Fifteenth Symposium, ASTM STP 833*, R. J. Sanford, Ed., American Society for Testing and Materials, 1984, pp. 577–605.

[11] Orange, T. W., "Method for Estimating Crack Extension Resistance Curve From Residual-Strength Data," NASA TP-1753, National Aeronautics and Space Administration, Nov. 1980.

[12] Newman, J. C., Jr., and Mall, S., "Plastic Zone Size and CTOD Equations for the Compact Specimen," *International Journal of Fracture*, Vol. 24, 1984, pp. R59–R63.

[13] McCabe, D. E., "Data Development for ASTM E24.06.02 Round Robin Program on Instability Prediction," NASA CR-159103, National Aeronautics and Space Administration, Aug. 1979.

[14] Newman, J. C., Jr., this volume, pp. 5–96.

[15] Newman, J. C., Jr., "An Improved Method of Collocation for the Stress Analysis of Cracked Plates With Various Shaped Boundaries," NASA TN D-6376, National Aeronautics and Space Administration, Aug. 1971.

[16] Newman, J. C., Jr., "Crack-Opening Displacements in Center-Crack, Compact, and Crack-Line-Wedge-Loaded Specimens," NASA TN D-8268, National Aeronautics and Space Administration, July 1976.

[17] Smith, E., "Fracture at Stress Concentrations" in *Proceedings*, First International Conference on Fracture, Sendai, Japan, 1965, pp. 139–151.

[18] Tada, H., Paris, P. C., and Irwin, G. R., *The Stress Analysis of Cracks Handbook*, Del Research Corp., Hellertown, PA, 1973.

Summary

Summary

The papers in this publication are divided into two major sections: (1) an experimental and predictive round robin, and (2) the presentation of four elastic-plastic fracture criteria. The fracture criteria are used to predict the failure of flawed metallic structures under elastic-plastic conditions. The failure predictions are based upon theory, coupled with critical material parameters which are measured from laboratory fracture specimens. Each method describes the steps required for its application, and sample calculations are included. The results of a round robin are also discussed in which these and other methods were used to predict failure loads for cracked structural configurations based on data from compact specimens. By combining various predictive methods into one volume, a reference basis is provided to judge the performance of these methods and to assess their advantages as well as their limitations. It is hoped that the combined presentation of several methods will provide a basis for their improvement and possible consolidation.

Experimental and Predictive Round Robin

A round robin on fracture was conducted by ASTM Task Group E24.06.02 on Application of Fracture Analysis Methods. The objective of the round robin was to verify whether fracture analysis methods currently used could predict failure loads on complex structural components containing cracks. Results of fracture tests conducted on various-size compact specimens made of 7075-T651 aluminum alloy, 2024-T351 aluminum alloy, and 304 stainless steel were supplied as baseline data to 18 participants. These participants used 13 different methods to predict failure loads on other compact specimens, middle-crack tension specimens, and structurally configured specimens.

The methods used in the round robin included: linear-elastic fracture mechanics corrected for size effects or for plastic yielding, Equivalent Energy, the Two-Parameter Fracture Criterion (TPFC), the Deformation Plasticity Failure Assessment Diagram (DPFAD), the Theory of Ductile Fracture, the K_R-curve with the Dugdale model, an effective K_R-curve, derived from residual strength data, the effective K_R-curve, the effective K_R-curve with a limit-load condition, limit-load analyses, a two-dimensional finite-element analysis using a critical crack-tip-opening displacement (CTOD) criterion with stable crack growth, and a three-dimensional finite-element analysis using a critical crack-front singularity parameter with a stationary crack. The failure loads were unknown to all participants except one of the task group chairman, who used one of the TPFC applications and the critical CTOD criterion.

For 7075-T651 aluminum alloy, the best methods (predictions within 20% of experimental failure loads) were: the effective K_R-curve, the critical CTOD criterion using a finite-element analysis, and the K_R-curve with the Dugdale model. For the 2024-T351 aluminum alloy, the best methods were: the TPFC, the critical CTOD criterion, the K_R-curve with the Dugdale model, the DPFAD, and the effective K_R-curve with a limit-load condition. For 304 stainless steel, the best methods were: the limit load (or plastic collapse) analyses, the critical CTOD criterion, the TPFC, and the DPFAD.

In conclusion, many of the fracture analysis methods tried could predict failure loads on various crack configurations for a wide range in material behavior. In several cases, the analyst had to select the method he thought would work the best. This would require experience and engineering judgment. Some methods, however, could be applied to all crack configurations and materials considered. Many of the large errors in predicting failure loads were due to improper application of the method or human error. As a result of the round robin, many improvements have been made in these and other fracture analysis methods.

Elastic-Plastic Fracture Mechanics Methodology

The K_R-curve method described by McCabe and Schwalbe uses as its basis the elastic-plastic resistance curve defined by ASTM Recommended Practice on R-curve Determination (E 561) to predict instability in a structure or specimen. The predictive capability is restricted to those cases where the specimen or component is stressed below net-section yield. The K_R-curve is a modified linear-elastic approach that has been extended to handle elastic-plastic crack-tip field conditions. An equivalence exists between K_R and J_R to the point of maximum load (bend configurations) and the approach is not different from the J_R prediction methodology in this region of equivalence. By eliminating elastic-plastic deformation requirements, the K_R method provides a simple approach to treat complex configurations. Instability can be predicted for any configuration for which a linear-elastic K_I analysis exists. Both the conditions of load control and displacement control are treated. The paper outlines the computational steps, and its application is illustrated with three example problems. The method has been used for ultra-high-strength sheet materials; certain restrictions apply for more-ductile materials.

Bloom presents a DPFAD to assess the integrity of a flawed structure. The approach is similar to the R-6 Failure Assessment Diagram developed by the Central Electricity Generating Board in the United Kingdom. This is a simple engineering procedure for the prediction of instability loads in flawed structures, which uses deformation plasticity, the J-integral estimation scheme, and handbook solutions. The DPFAD is broad-based in that it treats both brittle fracture and net-section plastic collapse. A failure assessment curve is defined in terms of stress-intensity-factor-to-fracture-toughness ratio against applied-stress-to-net-section-plastic-collapse-stress ratio. An assessment point is considered to be safe or unsafe based upon its position in the DPFAD. The method addresses ductile tearing by redefining the failure assessment curve as the boundary between stable and unstable crack growth. The method requires a fully plastic solution for flawed structures of interest. In addition, the amount of stable crack growth permitted in the analysis could be small in that the limits of J-controlled growth must be satisfied.

Ernst and Landes describe a failure prediction method based upon a modified $J(J_M)$-resistance curve. The method requires an experimentally determined J_M-resistance curve and two calibration functions that relate load, load-point displacement, crack length and J_M for the configuration of interest. An elastic-plastic analysis for J_M for the flawed structure of interest is required. The method enables one to compute the maximum load or instability load for load-controlled conditions and the entire load-load point displacement of the untested structure. Instability can also be computed using the J_M-T_M diagram where T_M is the tearing modulus of the material. The J_M parameter is different from the J-integral value computed from deformation theory (J_D). Specifically, J_M is no longer a path-independent integral. On the other hand, J_M appears to allow for crack extension far in excess of that permitted by J_D, thereby, providing a potentially superior parameter for flawed structural characterization. For the method to be applicable, both the crack growth mechanism and mechanical constraint must be the same in the structure as in the specimen used to obtain the J_M-resistance curve. In addition, this procedure does not treat cases where brittle (cleavage) failure may occur in structural steels.

In the V_R-curve method described by Newman, the crack growth resistance to fracture is expressed in terms of crack-tip-opening displacement. Basically, the V_R curve method is quite similar to the K_R or J_R methods, except that the "crack drive" is written in terms of displacement instead of K or J. Unlike the K_R and J_R methods, however, the V_R-curve method cannot be applied for crack extensions beyond maximum load. The reason for this behavior was not given. A relationship between crack-tip-opening displacement, crack length, specimen type, and tensile properties is derived from the Dugdale model. Because the Dugdale model is obtained from superposition of two elastic crack problems, the V_R-curve method can be applied to any crack configuration for which these two elastic solutions have been obtained. The method requires an experimentally determined V_R-resistance curve on the material of interest. The V_R-curve can be determined from either load-crack extension data or from failure load data using the initial crack length. In the latter method, no crack extension data are required. Thus, fracture tests conducted 20 to 30 years ago can be used to obtain the V_R-curve. The analysis procedures used to predict stable crack growth and instability of any through-the-thickness crack configuration made of the same material and thickness, and tested under the same environmental conditions, are presented. Three example calculations and predictions are shown. The various limitations of the method are also given.

J. C. Newman, Jr.
NASA Langley Research Center, Hampton, VA
23665; task group co-chairman and editor.

F. J. Loss
Materials Engineering Associates, Inc., Lanham,
MD 20706; task group co-chairman and editor.

Author Index

Subject Index